solutions@syngress.com

With more than 1,500,000 copies of our MCSE, MCSD, CompTIA, and Cisco study guides in print, we continue to look for ways we can better serve the information needs of our readers. One way we do that is by listening.

Readers like yourself have been telling us they want an Internet-based service that would extend and enhance the value of our books. Based on reader feedback and our own strategic plan, we have created a Web site that we hope will exceed your expectations.

Solutions@syngress.com is an interactive treasure trove of useful information focusing on our book topics and related technologies. The site offers the following features:

- One-year warranty against content obsolescence due to vendor product upgrades. You can access online updates for any affected chapters.

- "Ask the Author" customer query forms that enable you to post questions to our authors and editors.

- Exclusive monthly mailings in which our experts provide answers to reader queries and clear explanations of complex material.

- Regularly updated links to sites specially selected by our editors for readers desiring additional reliable information on key topics.

Best of all, the book you're now holding is your key to this amazing site. Just go to **www.syngress.com/solutions**, and keep this book handy when you register to verify your purchase.

Thank you for giving us the opportunity to serve your needs. And be sure to let us know if there's anything else we can do to help you get the maximum value from your investment. We're listening.

www.syngress.com/solutions

SYNGRESS®

1 YEAR UPGRADE
BUYER PROTECTION PLAN

SYNGRESS.COM
1 YEAR UPGRADE
BUYER PROTECTION PLAN

10 COOL LEGO® Mindstorms™

DARK SIDE ROBOTS, TRANSPORTS, AND CREATURES

Kevin Clague
Søren Rolighed
Miguel Agullo
Hideaki Yabuki

KEY	SERIAL NUMBER
001	T945G3ERT5
002	T4K9PLDLE4
003	L3N5SPT64N
004	B39SH5J74N
005	U6N6Y3KM7H
006	N2VN9G4RC7
007	TU8B2BVH3W
008	9R5MQPBR5A
009	M483N546HA
010	H2Q7SGT6YS

PUBLISHED BY
Syngress Publishing, Inc.
800 Hingham Street
Rockland, MA 02370

10 Cool LEGO MINDSTORMS Dark Side Robots

Printed in the United States of America

1 2 3 4 5 6 7 8 9 0

ISBN: 1-931836-59-0

Acquisitions Editor: Jonathan Babcock
Copy Editor: Kate Glennon

Cover Designer: Michael Kavish
Page Layout and Art by: Shannon Tozier

Distributed by Publishers Group West in the United States and Jaguar Book Group in Canada.

Syngress Acknowledgments

We would like to acknowledge the following people for their kindness and support in making this book possible.

A special thanks to Matt Gerber at Brickswest for his help and support for our books.

Karen Cross, Lance Tilford, Meaghan Cunningham, Kim Wylie, Harry Kirchner, Kevin Votel, Kent Anderson, Frida Yara, Jon Mayes, John Mesjak, Peg O'Donnell, Sandra Patterson, Betty Redmond, Roy Remer, Ron Shapiro, Patricia Kelly, Andrea Tetrick, Jennifer Pascal, Doug Reil, David Dahl, Janis Carpenter, and Susan Fryer of Publishers Group West for sharing their incredible marketing experience and expertise.

Jacquie Shanahan, AnnHelen Lindeholm, David Burton, Febea Marinetti, and Rosie Moss of Elsevier Science for making certain that our vision remains worldwide in scope.

David Buckland, Wendi Wong, Marie Chieng, Lucy Chong, Leslie Lim, Audrey Gan, and Joseph Chan of Transquest Publishers for the enthusiasm with which they receive our books.

Kwon Sung June at Acorn Publishing for his support.

Jackie Gross, Gayle Voycey, Alexia Penny, Anik Robitaille, Craig Siddall, Darlene Morrow, Iolanda Miller, Jane Mackay, and Marie Skelly at Jackie Gross & Associates for all their help and enthusiasm representing our product in Canada.

Lois Fraser, Connie McMenemy, Shannon Russell, and the rest of the great folks at Jaguar Book Group for their help with distribution of Syngress books in Canada.

A special welcome to the folks at Woodslane in Australia! Thank you to David Scott and everyone there as we start selling Syngress titles through Woodslane in Australia, New Zealand, Papua New Guinea, Fiji Tonga, Solomon Islands, and the Cook Islands.

The Syngress Production Department would like to thank Vesna Williams, Sasha Williams, and Kevin Cawley for all of their hard work on this project.

Contributors

Hideaki Yabuki works as a Media Activist promoting new technologies to the next generation. To him, robotics is the most important of these technologies. He was first introduced to LEGO robots in 1985 by a friend of his, Mr. Kurita, who had recently returned from the MIT Media Lab with some LEGO Dacta products. Influenced also by Dr. Seymour Papert's book, *Mindstorms: Children, Computers, and Powerful Ideas*, Hideaki feels that LEGOs offer a hands-on approach to learning that is often missing these days in our digital world. His robot in this book, the Scorpion, is the result of much trial and error on his part. Hideaki would like to thank J.P. Brown, without whom he could not have become involved with this book. He would also like to thank Kevin Clague, inventor of the LPub application for semi-automated LEGO instruction creation. Lastly, Hideaki would like to give his deepest thanks for the support of his mother, Rei, and his dear wife and son, Keiko and Kei. Hideaki has a dream that one day the people of Japan will be able to join hands with biped robots as friends.

Kevin Clague graduated in 1983 from Iowa State University with a bachelor's of Science degree in Computer Engineering. For the past 18 years, Kevin has worked as a Diagnostic Engineer at the Amdahl Corporation. For the last two years, he has also acted as a Senior Staff Engineer doing verification work at Sun Microsystems on their Ultra-Sparc V RISC processor. Kevin has two major hobbies: theatrical lighting and LEGO MINDSTORMS. Kevin has been playing with the RIS 1.5 for several years now and is currently working on LPub, an application to revolutionize the world of creating online LEGO building instructions.

Miguel Agullo was born in Spain but has lived abroad for long periods of time, from the Far East to South America, from central Europe to the U.S. His wide range of interests is responsible for his work in such diverse industries as finance, media, aeronautics, and antique trading. Trained as a journalist and impressed with the candor and resourcefulness of the online LEGO community, Miguel tries to give something back by regularly updating his Web site at www.geocities.com/technicpuppy with instructions for new models, new Ldraw pieces, and anything he thinks is worth sharing with other LEGO aficionados. His building interests revolve around robotics, and specifically biomechanics: creating mechanisms that mimic the behavior of natural devices such as legs or arms. His creations include biped walkers, robots that jump, and a fully functional (including a brake!) LEGO motorcycle. His current hobbies include boating, biking, traveling and learning Thai and Dutch (his wife is Thai and they live in Amsterdam, which explains the bike and the boat).

Søren Rolighed is a data warehouse consultant, working on building and maintaining databases for telco-data in the largest data warehouse in Denmark. Like almost all Danish kids, he started playing with LEGO at an early age. As an adult he has continued with his passion for LEGOs, and the introduction of the LEGO Technics and LEGO MINDSTORMS kits opened up a whole new world of possibilities! Søren enjoys building LEGO Mindstorm robots and programming them with his computer. He would like to thank his three great children, Emma, Oscar, and Victor, for their patience and understanding when dad has to requisition all of their interesting motors, gears, and special parts for his own LEGO creations. Søren and his family live in Copenhagen, Denmark.

J.P. Brown is a Consultant Environmental Conservator who has worked on such historical sites as Independence Hall, Philadelphia, PA and George Washington's mansion, Mount Vernon, VA. He first became interested in LEGO MINDSTORMS in July 1999, but his interest did not really take off until he discovered Dave Baum's *Not Quite C* (NQC) programming environment for the RCX brick later that year. He quickly became involved as a moderator for LEGO MINDSTORMS forums on the Web, and was later selected by LEGO as a preview builder for the MINDSTORMS Vision Command system (along with Søren Rolighed and several others). His robot, Biped II, won the February 2001 MINDSTORMS Hall of Fame, Special Competition, but he is perhaps best known for his Rubik's Cube solving robot, CubeSolver, which was featured in the *New York Times* in October 2001 and other papers around the world.

Contents

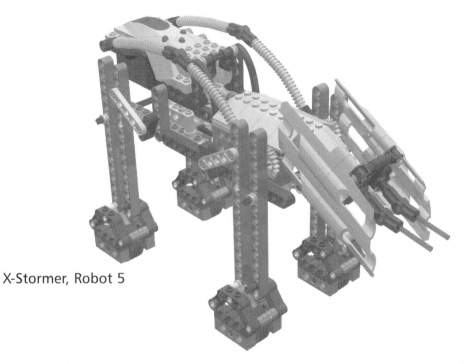

X-Stormer, Robot 5

Foreword

This book is the result of a lifetime exploring the world of LEGO. As a kid, I loved taking mechanical things apart to figure out how they worked, then either putting them back together or using the parts to invent something new, so I really enjoyed playing with LEGOs–despite the fact that the things I built often fell apart! Eventually, though, I grew out of them. Several years after I grew out of this first LEGO phase, LEGO introduced the TECHNIC line of LEGOs, a superior set of building parts allowing sturdy structures and complicated designs with gears, levers, and all kinds of moving mechanics, and then introduced the MINDSTORMS line. Two and a half years ago, my wife got me the LEGO MINDSTORMS Dark Side Developer Kit for Christmas–I immediately built the AT-AT walker and was hooked on LEGOs all over again. A few months later, I went out and got the Robot Invention System 1.5, and thus began my second LEGO phase. LEGO's TECHNIC and MINDSTORMS kits have allowed me to pursue both my passion for building with LEGOs and my interest in all things mechanical.

This book is a first for my co-authors, Syngress, and myself. I've really enjoyed working with and getting to know the other authors (and the people at Syngress) during the development of this book. The Internet has allowed Syngress to gather a team of authors from many different countries, including Britain, Holland, Denmark, Japan, and the United States. Before I began working on the book, I'd met J.P. Brown in person when he was doing some consulting work in St. Paul, Minnesota. I'd also had some contact with Miguel Agullo (a.k.a. "Technic Puppy") because he runs a Web page dedicated specifically to LEGO biped robots, which are my favorite type of robot to invent. I'd heard of Soren Rolighed because he made it into the LEGO MINDSTORMS Hall of Fame, and I'd also heard of Hideaki Yabuki (a.k.a. "Joda", which is

Japanese for Yoda) because I saw one of his fabulous grabber arms featured on the LEGO MINDSTORMS Web site.

When Jonathan Babcock at Syngress first contacted me about working on *10 Cool LEGO MINDSTORMS Dark Side Robots, Transports, and Creatures*, he told me it was going to be a step-by-step building instruction book. I was curious how we were going to lay out all the artwork for the book, and Syngress didn't initially have a firm plan on how this would be done. After talking it out with Jonathan, it was decided that I would write some tools to help do the job. Many tools had already been developed by others and given away over the Internet for the LEGO community to share. I know that without those tools this book would never have made it to press as polished and as quickly as it has.

The groundwork for all the tools used was laid out by the late James Jessiman. James created the LDraw program on which all the other tools are based. MLCad, written by Michael Lachmann, is a second-generation computer aided design (CAD) program that is used to enter robotic designs. The third tool that we use is L3P, written by Lars C. Hassing. L3P converts LDraw files to POV-Ray files. POV-Ray is the program that produces the final images that you see in this book. POV-Ray has a long history of contributors. The Windows version we used was coordinated by Chris Cason.

I wrote a program called LPub for this book that combines all these tools together to generate the step-by-step construction images and the corresponding part-list images. The development of LPub has been an adventure in and of itself, and has kept me very busy over the past several months. I will be cleaning it up, making it more robust, and making it available for all to use in the near future.

I hope you enjoy building the robots in this book—some very creative minds have done some amazing work inventing the robots you'll find here. We encourage you to use your own creativity to modify our designs after you've built some of them, in the true LEGO building tradition.

If you like this book—and I think you will—be sure to look for two similar books that Syngress will be putting out in the near future: *10 Cool LEGO MINDSTORMS Robotics Invention System 2.0 Projects: Amazing Projects You Can Build in Under an Hour* (ISBN 1-931836-61-2) and *10 Cool LEGO MIND-STORMS Ultimate Builders Projects: Amazing Projects You Can Build in Under an Hour* (ISBN 1-931836-60-4).

–Kevin L. Clague

Robot 1

Super Battle Droid

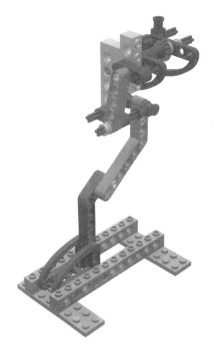

In *Episode II: Attack of the Clones*, the evil Sith lord Darth Sidious headed a separatist movement in the Republic, with the nefarious goal of ruling the entire galaxy. Darth Sideous had turned Jedi Knight Count Dooku to the dark side, and used him to lead the separatist movement. The Techno Union, one of the members of the separatist movement, created an army of Super Battle Droids, which were more powerful and aggressive than the Battle Droids used by the Trade Federation in *Episode 1: The Phantom Menace*.

We first heard of the Clone Wars in *Episode IV: A New Hope*, when Luke Skywalker and old Ben Kenobi (Obi-Wan) were talking about Luke's father. In Episode II, on the planet Geonosis, we see the start of the Clone Wars, as the Jedi Knights wage war with the Battle Droid and Super Battle Droid army to rescue Senator Amadala, Jedi Knight Obi-Wan Kenobi, and his Padawan, Anakin Skywalker.

In this chapter you will build a LEGO Mindstorms version of a Super Battle Droid. It is a biped that walks on two legs by shifting its weight from side to side and moving its legs forwards and backwards. Building the Super Battle Droid can be tricky because the side-to-side weight shifting needs to be timed just right to match the legs going forwards and backwards.

Bricks & Chips...

Micro Scout Programs

Since Super Battle Droid can walk both forwards and in reverse, you can use any of the programs that come with the Micro Scout to power it.

The Body

The body is the heart of the Super Battle Droid. As such, we will construct it first, and then build each of the sub-assemblies that will be added to the body to create the finished Droid.

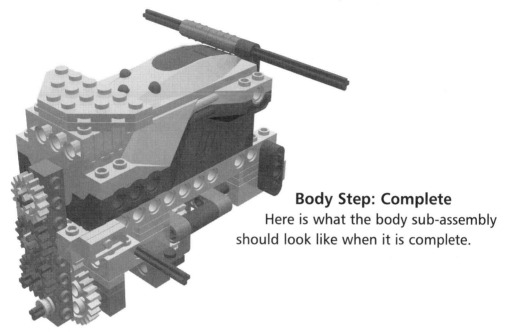

Body Step: Complete
Here is what the body sub-assembly should look like when it is complete.

Body Step: 1

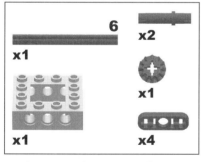

Make sure that two of the 3 holed struts (of which you are using 4 in this step) are pointing straight up. The two 3 holed struts on the opposite side should be pointing straight down (see picture).

Body Step: 2

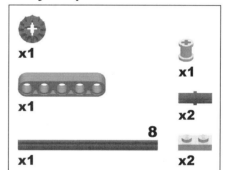

Body Step: 3

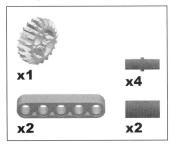

x1
x4
x2
x2

Make sure that the vertical 1x5 liftarm is attached to the axle through its middle hole (the third from either end).

Body Step: 4

x1
x2
10
x1
x1

When adding the axle and gears in **Body Step: 4**, make sure that the moving parts at the far end are positioned as shown in this image. It is also important to make sure the gears allow the #10 axle to be oriented as shown.

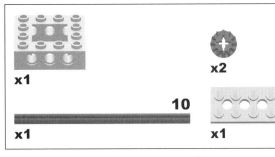

Body Step: 5

x2
x4
x1

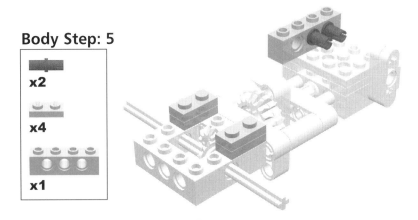

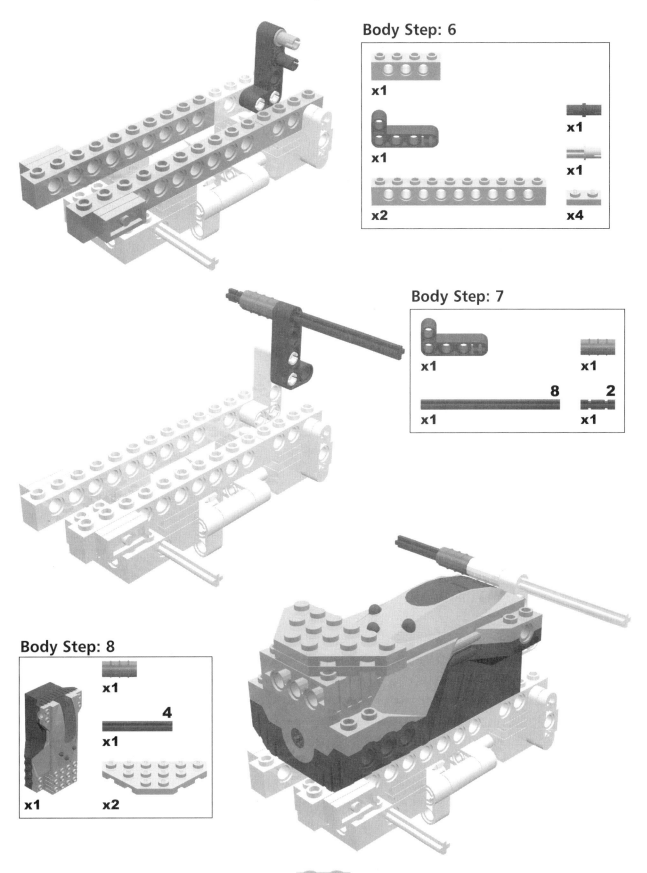

Body Step: 6

x1
x1
x2

x1
x1
x4

Body Step: 7

x1
x1

8
x1

2
x1

Body Step: 8

x1

4
x1

x1
x2

Body Step: 9

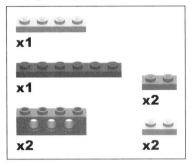

x1

x1

x2

x2

x2

Flip the body over so that the Micro Scout is on the bottom, and add the pieces as shown in **Body Step: 9**

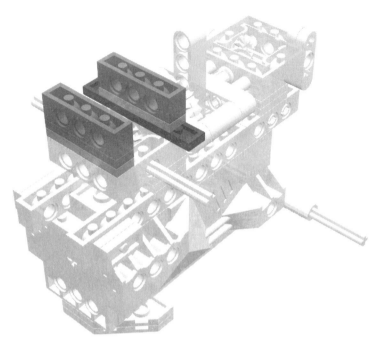

Body Step: 10

5

x1

x1

x1

x1

10

x1

x1

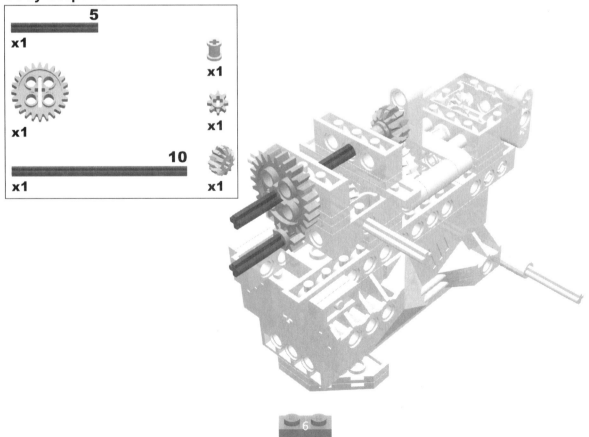

6

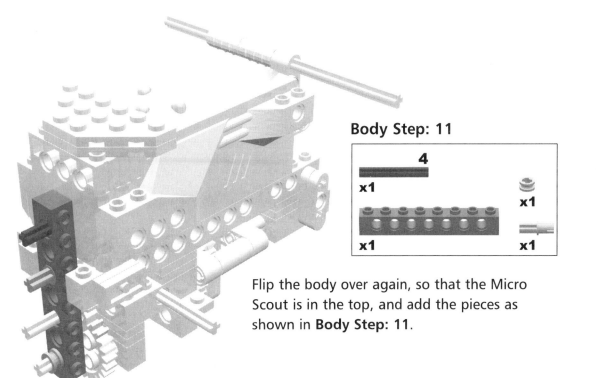

Body Step: 11

4	
x1	x1
x1	x1

Flip the body over again, so that the Micro Scout is in the top, and add the pieces as shown in **Body Step: 11**.

Body Step: 12

In **Body Step: 12**, you can use any three of the four 16 tooth gears that come with the Dark Side Developer's Kit. We have chosen to use two red and one gray.

x1

x2

The Right Leg

Right Leg Step: Complete

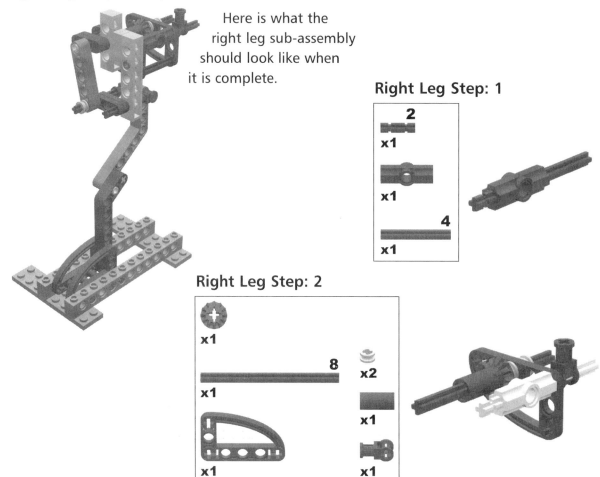

Here is what the right leg sub-assembly should look like when it is complete.

Right Leg Step: 1

2
x1

x1

4
x1

Right Leg Step: 2

x1

8
x1

x2

x1

x1

Right Leg Step: 3

x1   **4**
x1   x1

Right Leg Step: 4

x2

x1

x1

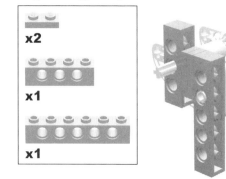

Right Leg Step: 5

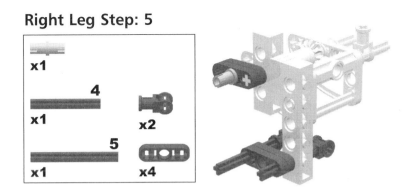

Right Leg Step: 6

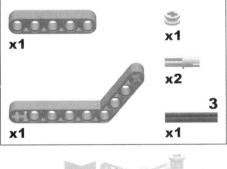

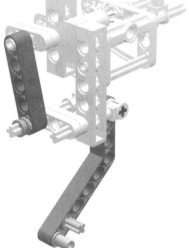

Right Leg Step: 7

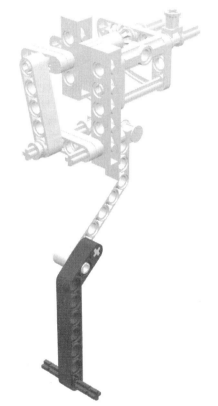

Right Leg Step: 8

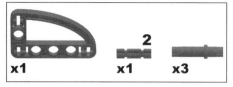

You will now begin construction of the foot itself.

Right Leg Step: 9

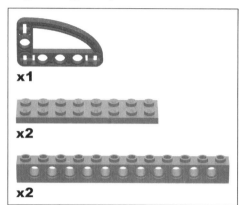

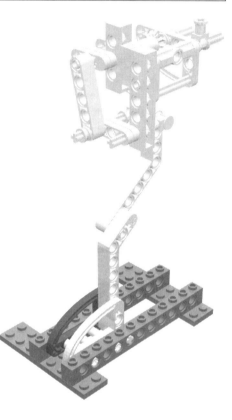

The Left Leg

Left Leg Step: Complete

Here is what the left
leg sub-assembly
should look like when
it is complete.

Left Leg Step: 1

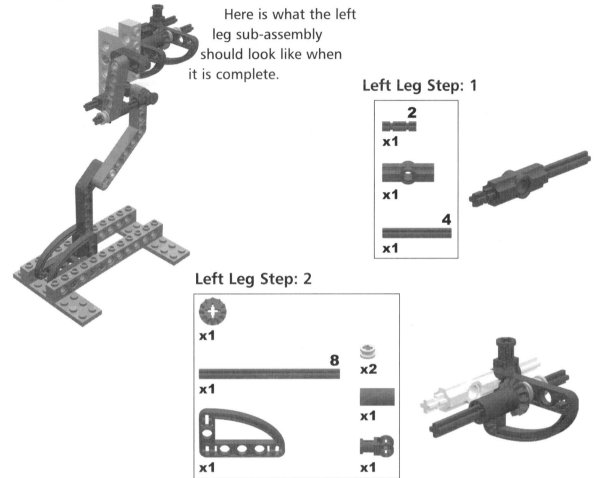

Left Leg Step: 2

Left Leg Step: 3

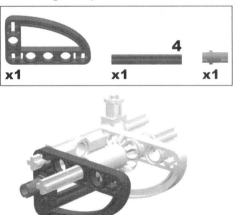

Left Leg Step: 4

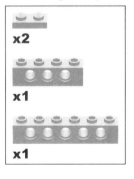

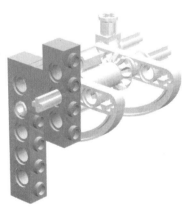

Left Leg Step: 5

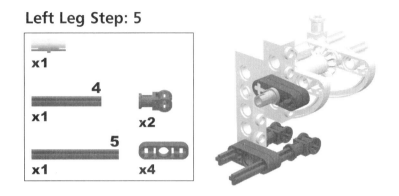

Left Leg Step: 6

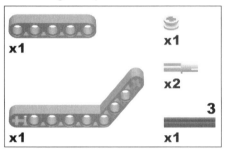

Left Leg Step: 7

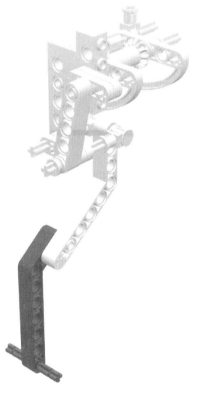

Left Leg Step: 8

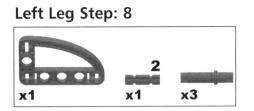

x1   x1   x3

You will now begin construction of the foot itself.

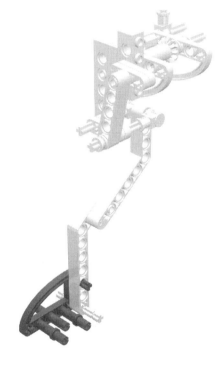

Left Leg Step: 9

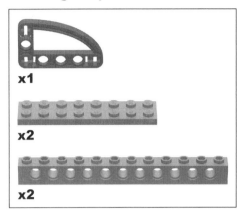

x1

x2

x2

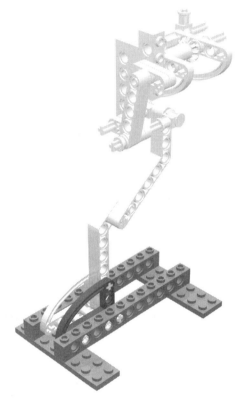

The Right Arm

Inventing…

Customizing the Arms

The arms are perhaps the easiest sub-assembly to customize on the Super Battle Droid. Using the pieces found in the Dark Side Developer's Kit and other LEGO kits, a wide variety of different possibilities for the arms are available. You could try building a sword, hands, or a variety of tools.

Right Arm Step: 1

x1   3
            x1

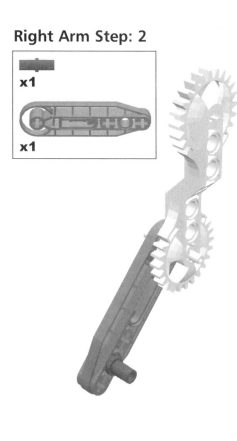

Right Arm Step: Complete

Here is what the right arm sub-assembly should look like when it is complete.

Right Arm Step: 3

x1   x1   R

Right Arm Step: 2

x1

x1

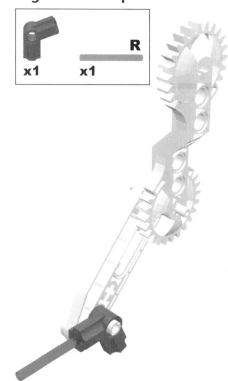

The Left Arm

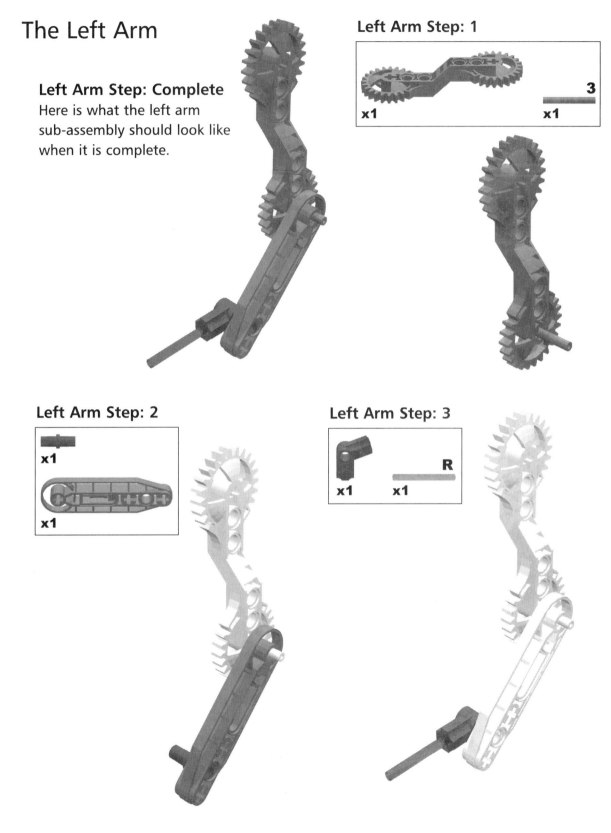

Left Arm Step: Complete
Here is what the left arm sub-assembly should look like when it is complete.

Left Arm Step: 1

x1   3   x1

Left Arm Step: 2

x1

x1

Left Arm Step: 3

x1   R   x1

The Chest

Chest Step: Complete

Here is what the chest sub-assembly should look like when it is complete.

Chest Step: 1

x2

x1

4

x1

Chest Step: 2

x1

x2

x1

Chest Step: 3

x1

x1

6

x1

x1

x2

x1

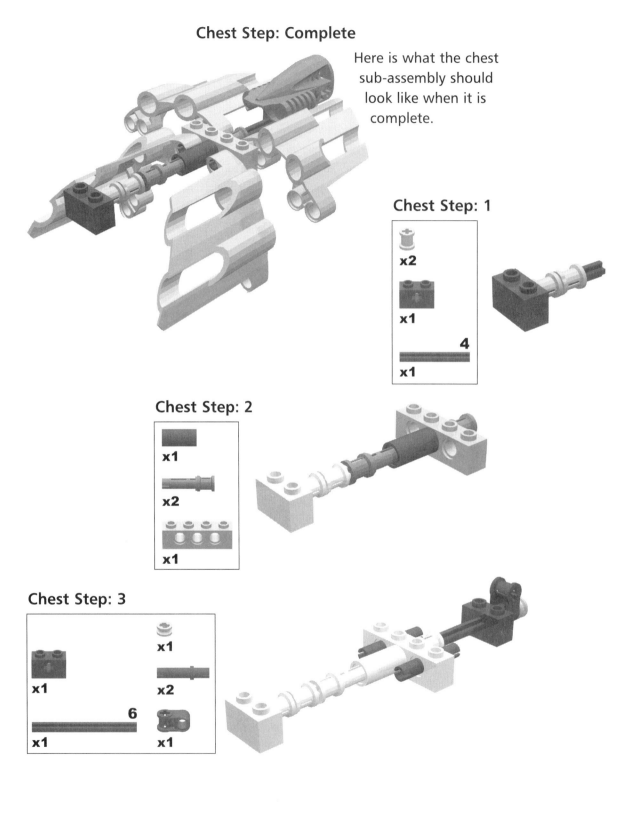

Chest Step: 4

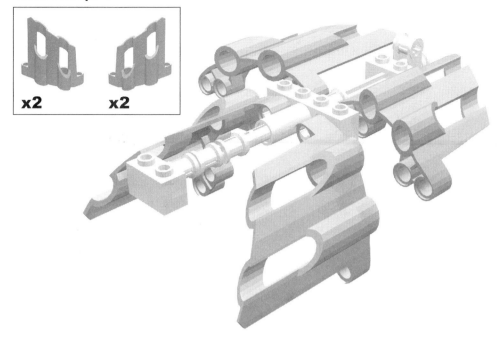

Chest Step: 5

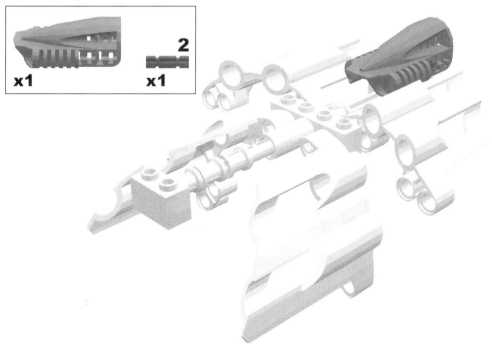

Final Assembly

You are now ready to take all of the sub-assemblies you have made (the right and left legs, the right and left arms, and the chest), and attach them to the body sub-assembly to create the finished Super Battle Droid.

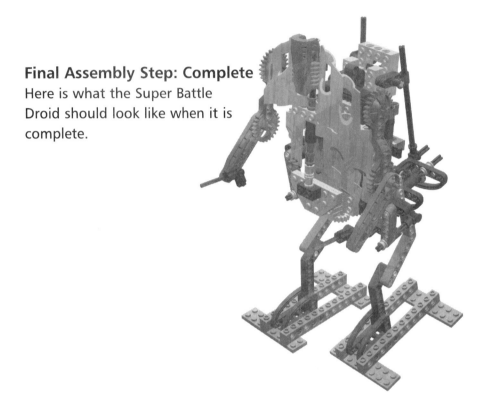

Final Assembly Step: Complete
Here is what the Super Battle Droid should look like when it is complete.

Final Assembly Step: 1
Begin the final assembly of the Super Battle Droid. Start by taking the body sub-assembly and holding it with the Micro Scout facing outwards, as shown in this picture.

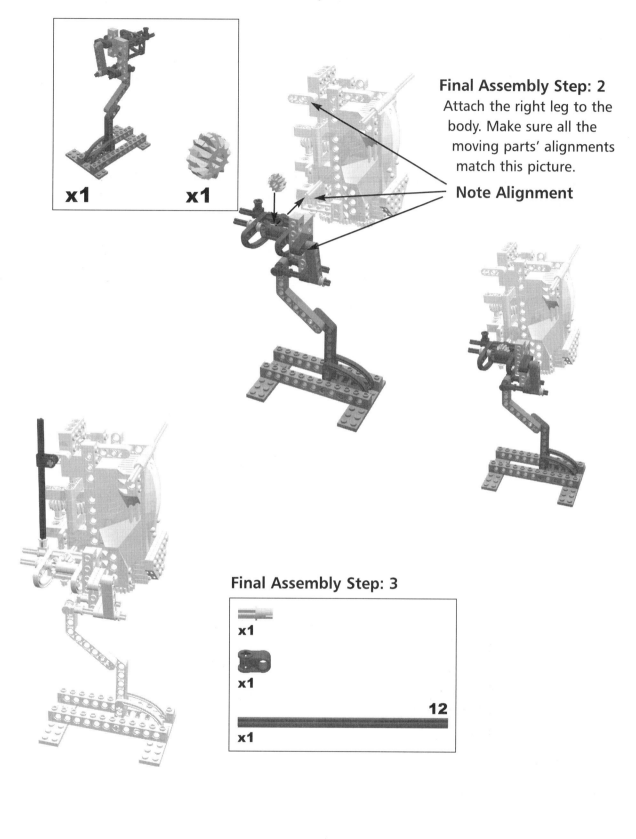

Final Assembly Step: 2
Attach the right leg to the body. Make sure all the moving parts' alignments match this picture.

Note Alignment

x1   x1

Final Assembly Step: 3

x1

x1

12

x1

Final Assembly Step: 4

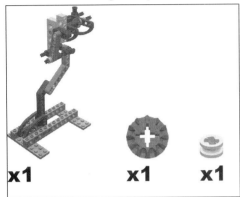

x1   x1   x1

Now, turn the body around and attach the left leg to the body. Since we are using the thinner, red 12t beveled gear for this leg, make sure that the small gray pulley goes behind the red gear, and that the gear itself meshes with the other red 12t beveled gear already attached to the leg. The gray pulley should be between the red gear you just added to the axle and the body of the Super Battle Droid. Make sure all the moving parts' alignments match this picture.

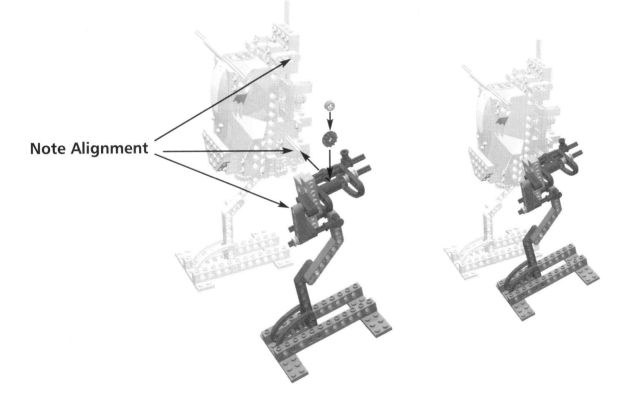

Note Alignment

Final Assembly Step: 5

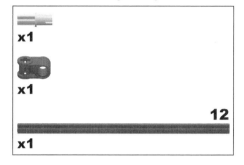

x1

x1

12

x1

Final Assembly Step: 6

Attach the arms, as shown. Make sure the guns on the ends of Super Battle Droid's arms are on the inside of the arms.

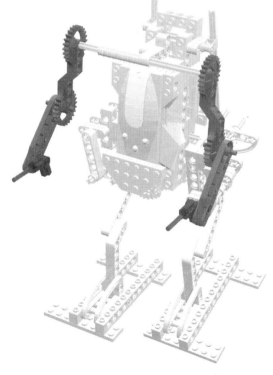

Final Assembly Step: 7

Lastly, attach the chest
sub-assembly to the
body, and Super Battle
Droid is complete!

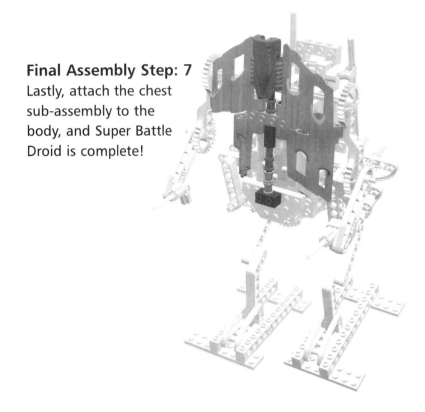

Robot 2

Orbital Defense Cannon

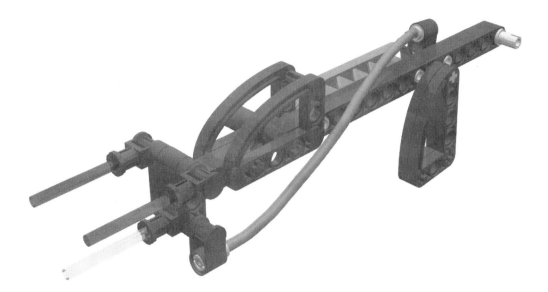

Orbital Defense Cannons are huge pieces of laser artillery that planets in the Alliance use to defend against bombardment from orbiting ships. Their dual laser cannons, guided by a targeting radar dish, can shoot down even the largest Imperial Star Destroyers and Corellian Cruisers. Many planets that can't afford to outfit a full fleet of starships choose instead to protect themselves from Imperial Starfleet attacks from space with Orbital Defense Cannons.

In order for the gears to work properly it is important that the Orbital Defense Cannon is sitting on a flat surface.

Bricks & Chips...

Micro Scout Programs

The Orbital Defense Cannon works best with Micro Scout programs #2 and #3. Program #2 will turn the Orbital Defense Cannon around in circles and "fire" its dual laser cannons. With program #3, the Orbital Defense Cannon will turn in circles and rotate its radar, "firing" its dual laser cannons when you trigger the Micro Scout with a light source (like a flashlight).

The Main Sub-Assembly

Main Step: Complete
Here is what the main sub-assembly should look like when it is complete.

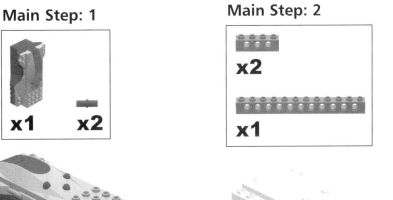

Main Step: 1

x1   **x2**

Main Step: 2

x2

x1

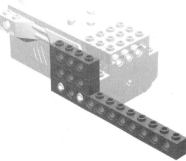

Main Step: 3

x2

Turn the main sub-assembly around and begin working on the opposite side.

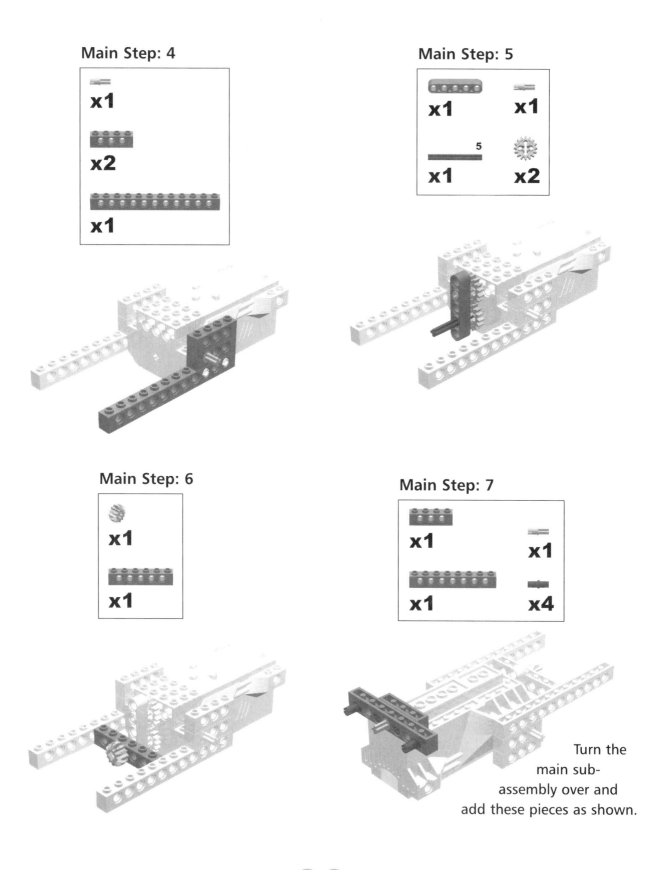

Main Step: 4

x1

x2

x1

Main Step: 5

x1   x1

x1 (5)   x2

Main Step: 6

x1

x1

Main Step: 7

x1   x1

x1   x4

Turn the main sub-assembly over and add these pieces as shown.

Main Step: 8

x8

x7

Main Step: 9

x4

x2

x2

Assemble
these pieces before
attaching them to the
main sub-assembly.

Main Step: 10

x2

3
x2

4
x2

Main Step: 11

x1

x2

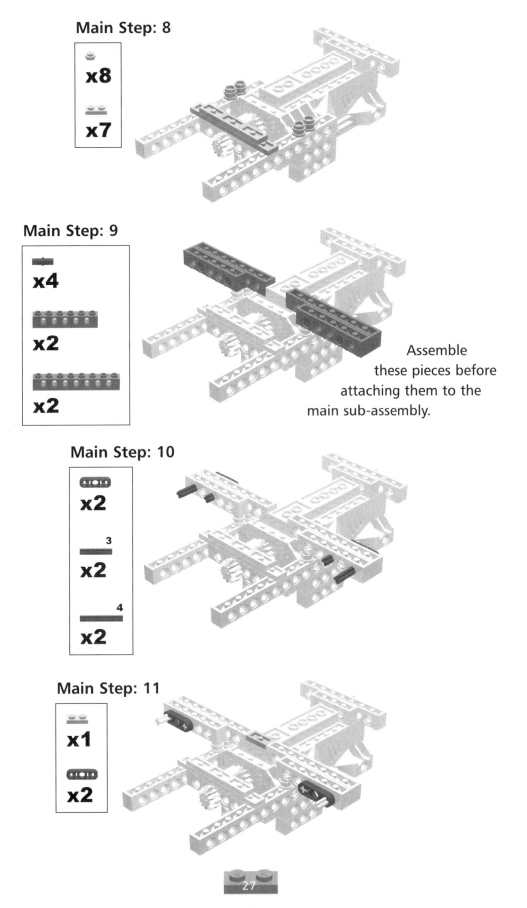

Main Step: 12

x1

x1

8

x1

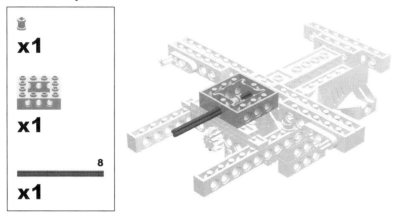

Main Step: 13

2

x2

x2

x1

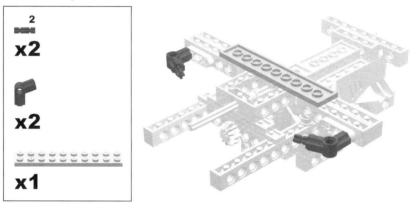

Main Step: 14

x2

x2

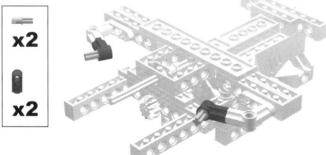

Main Step: 15

x2

x1

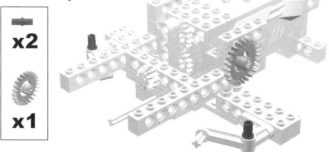

Main Step: 16

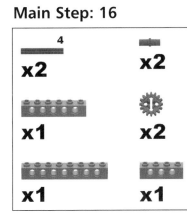

Assemble these pieces before attaching them to the main sub-assembly. Make sure that the 1x5 liftarm that projects from between the pieces you have just attached can move freely from side to side.

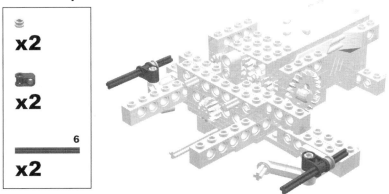

Main Step: 17

Main Step: 18

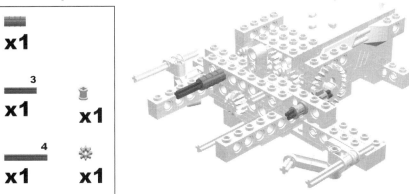

Main Step: 19

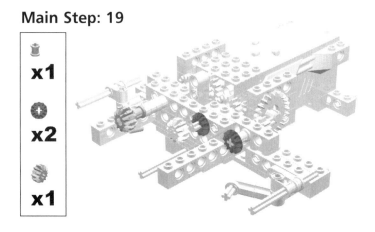

Main Step: 20

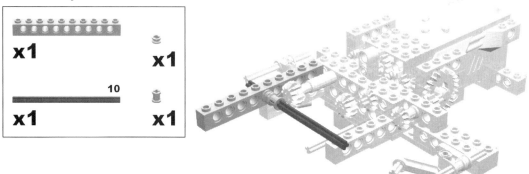

Main Step: 21

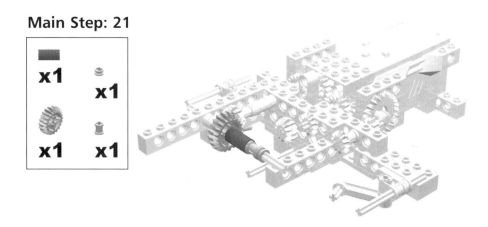

Main Step: 22

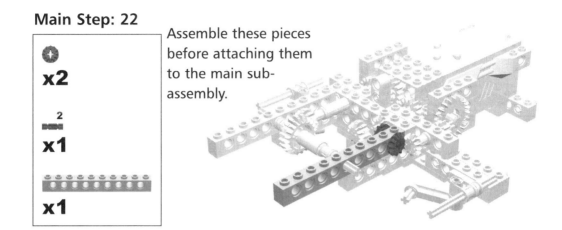

Assemble these pieces before attaching them to the main sub-assembly.

x2

2

x1

x1

Main Step: 23

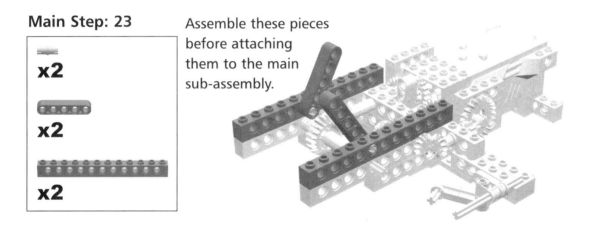

Assemble these pieces before attaching them to the main sub-assembly.

x2

x2

x2

Main Step: 24

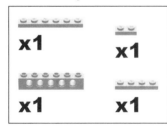

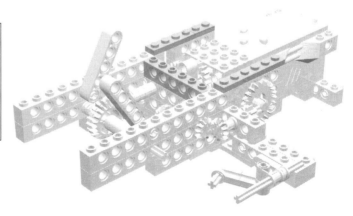

x1

x1

x1

x1

Main Step: 25

x2

x2

x1

x2

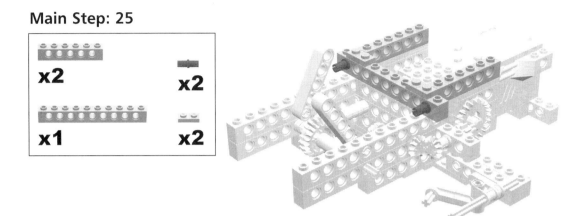

Main Step: 26

8
x1

x3

12
x1

x1

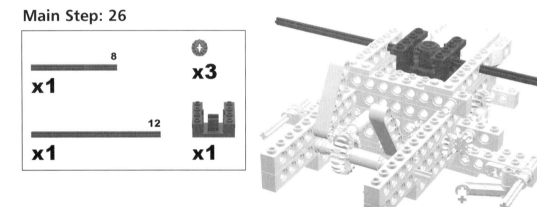

Main Step: 27

x1

3
x1

x1

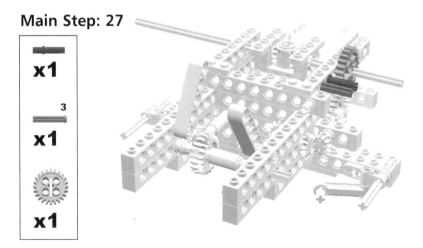

Main Step: 28

x3

4

x1

x1

x1

x1

x1

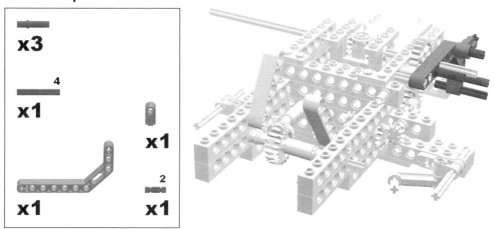

Main Step: 29

x1

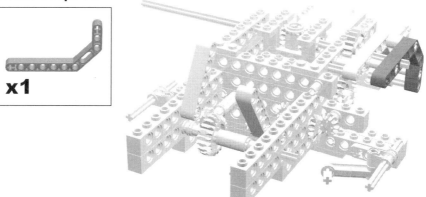

Main Step: 30

x2

3

x1

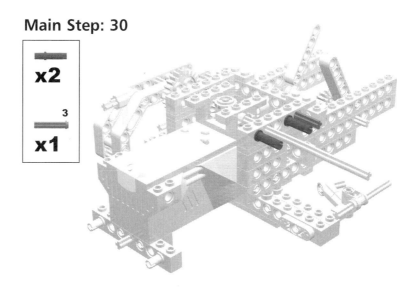

Main Step: 31

x3

4
x1

x1

x1

2
x1

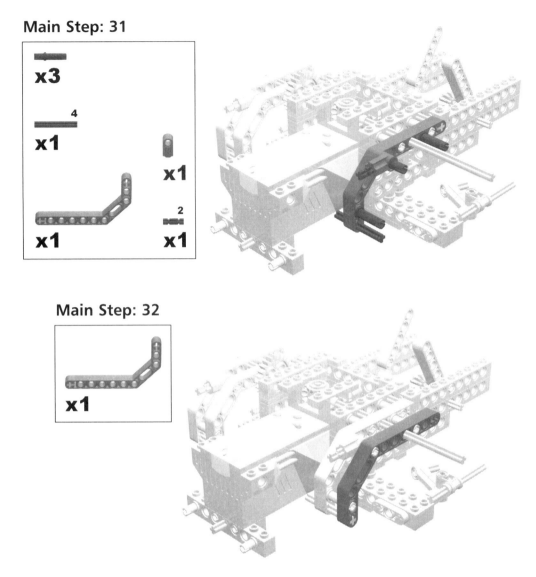

Main Step: 32

x1

Main Step: 33

x1

x6

x2

x1

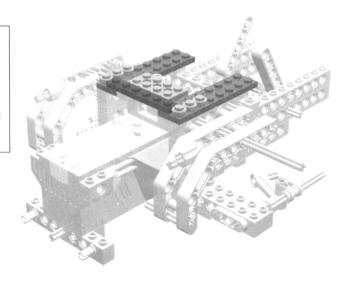

Main Step: 34

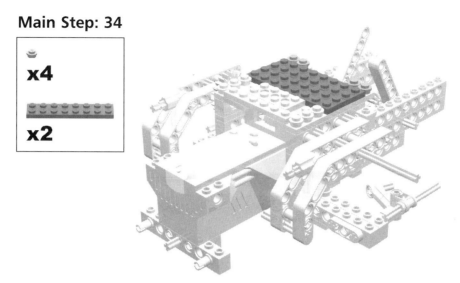

Main Step: 35

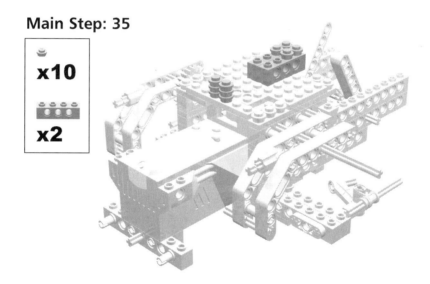

Main Step: 36

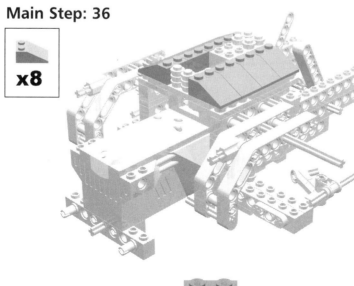

Main Step: 37

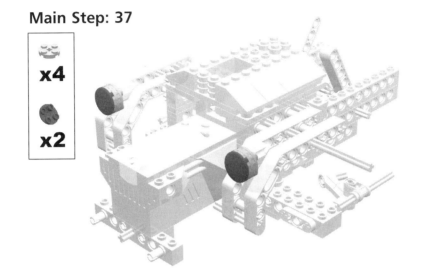

Main Step: 38

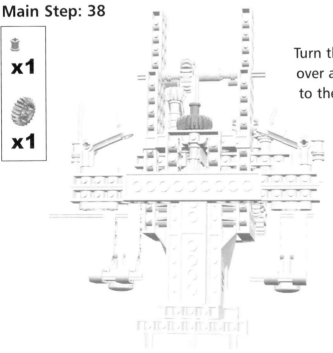

Turn the main sub-assembly over and add these pieces to the bottom as shown.

Main Step: 39

x4

x4

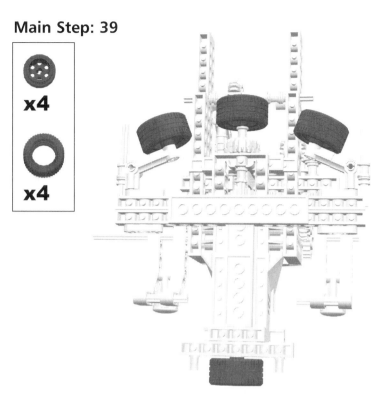

Main Step: 40

2

x2

x2

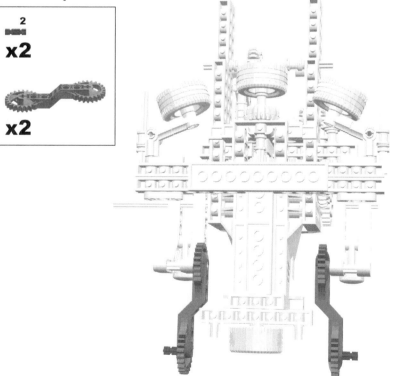

Main Step: 41

x4

x2

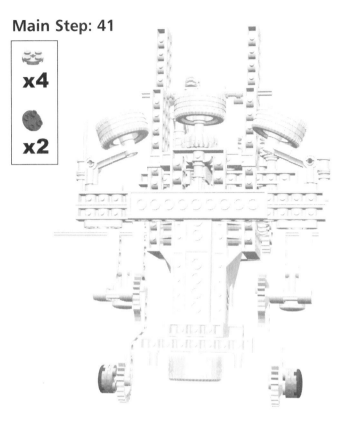

The Right Laser

Inventing...

Building Ideas

Try building a ship or robot that has a small penlight attached to it. This can serve as a target for the Orbital Defense Cannon. When the light from the penlight strikes the Micro Scout, the Orbital Defense Cannon will turn towards it and engage its dual laser cannons.

Right Laser Step: Complete

Here is what the right laser sub-assembly should look like when it is complete.

Right Laser Step: 1

Right Laser Step: 2

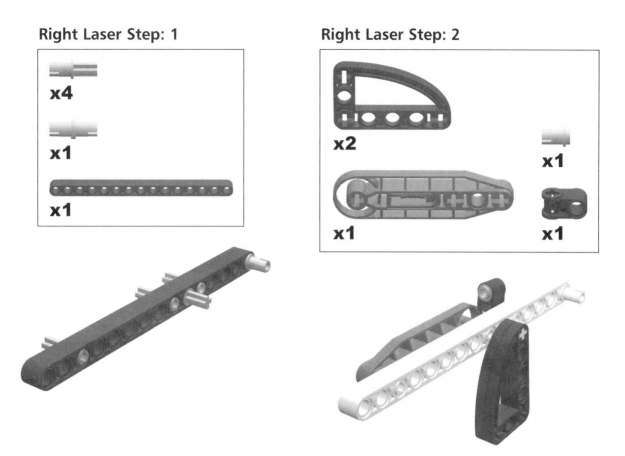

Right Laser Step: 3

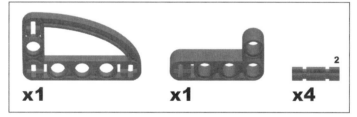

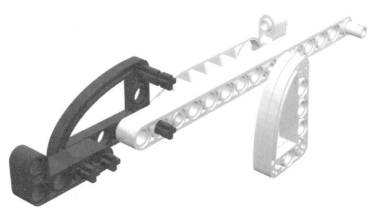

Right Laser Step: 4

x1 x1 x1

Right Laser Step: 5

x1

x1 ³ x1

x1 ⁴ x1

Right Laser Step: 6

x1 x1

x1 x2

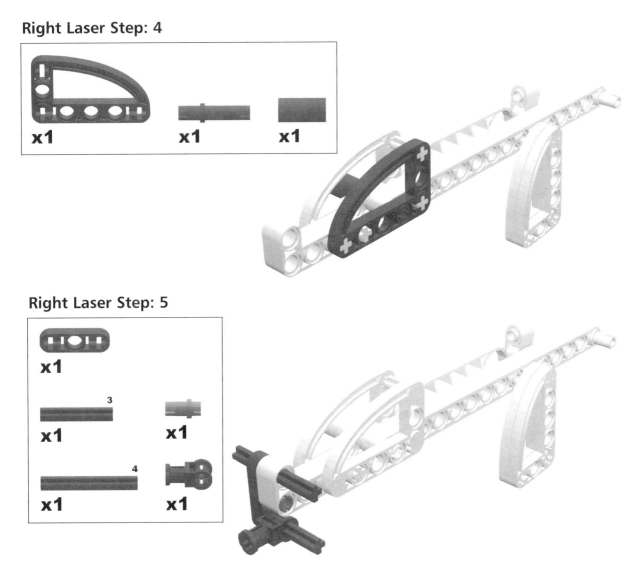

Right Laser Step: 7

x2

x1

We've used two red clear rods and one yellow clear rod in this step. You can use whatever colors you choose.

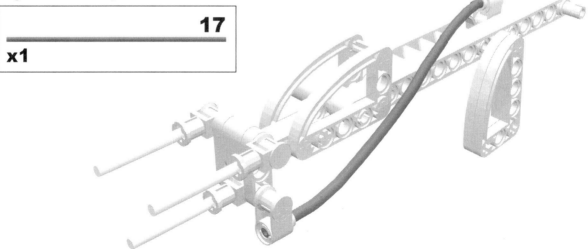

Right Laser Step: 8

17

x1

The Left Laser

Left Laser Step: Complete

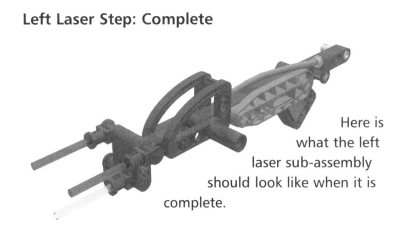

Here is what the left laser sub-assembly should look like when it is complete.

Left Laser Step: 1

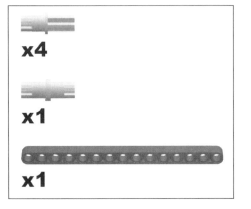

x4

x1

x1

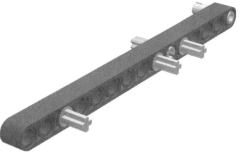

Left Laser Step: 2

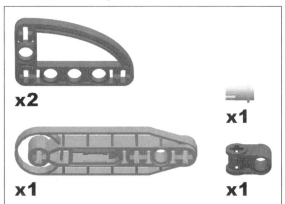

x2

x1

x1

x1

Left Laser Step: 3

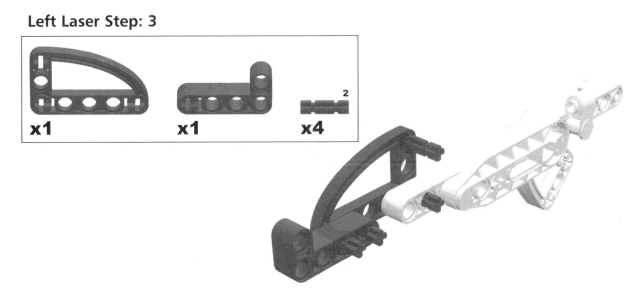

Left Laser Step: 4

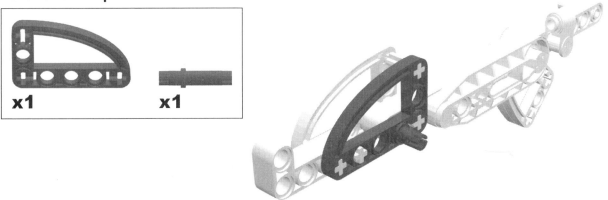

Left Laser Step: 5

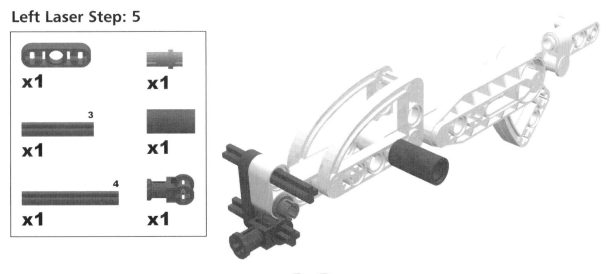

Left Laser Step: 6

x1   x1

x1   x2

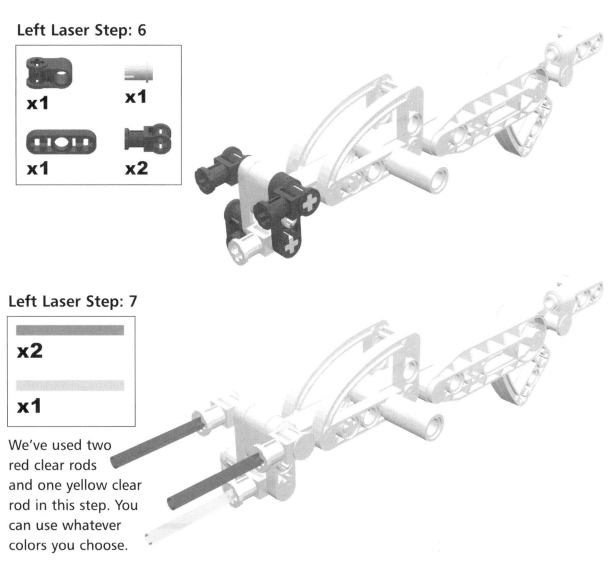

Left Laser Step: 7

x2

x1

We've used two red clear rods and one yellow clear rod in this step. You can use whatever colors you choose.

Left Laser Step: 8

17

x1

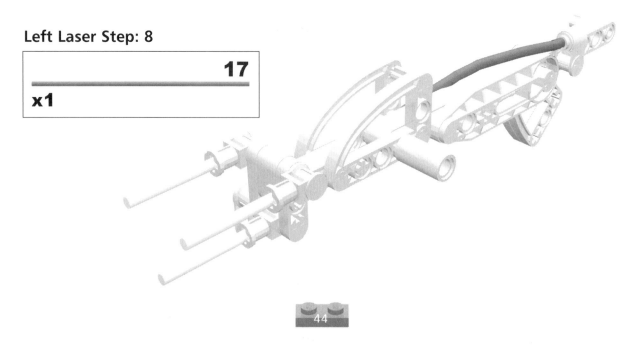

The Radar

Radar Assembly Step: Complete

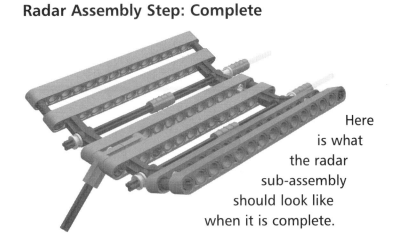

Here is what the radar sub-assembly should look like when it is complete.

Radar Assembly Step: 1

x1

2

x1

x1

Radar Assembly Step: 2

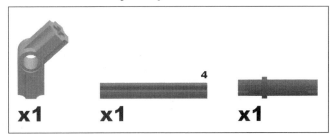

x1

4

x1

x1

Radar Assembly Step: 3

x4

x1

x2

Radar Assembly Step: 4

x4   x4

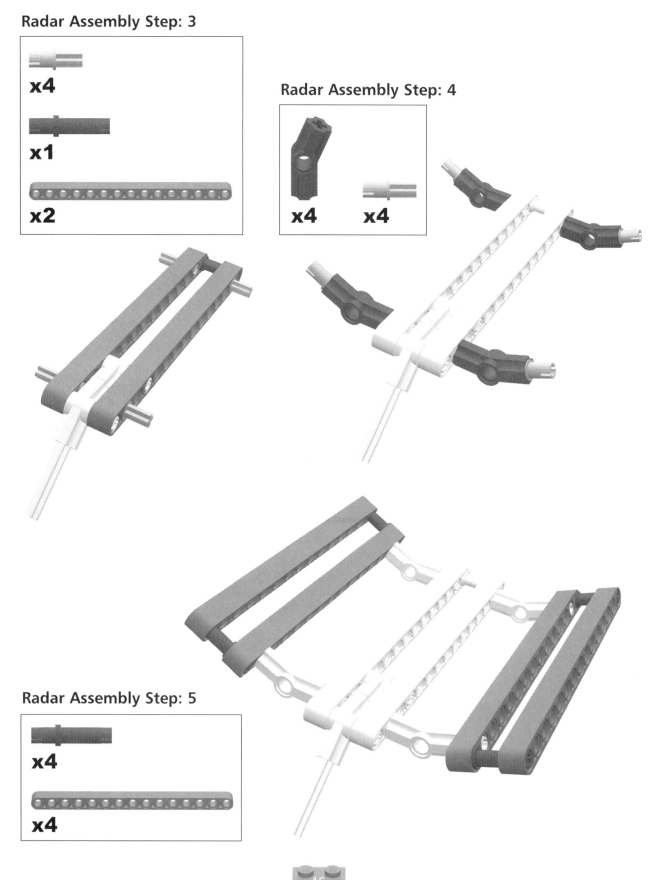

Radar Assembly Step: 5

x4

x4

Radar Assembly Step: 6

x2   **x2**

8
x4   **x2**

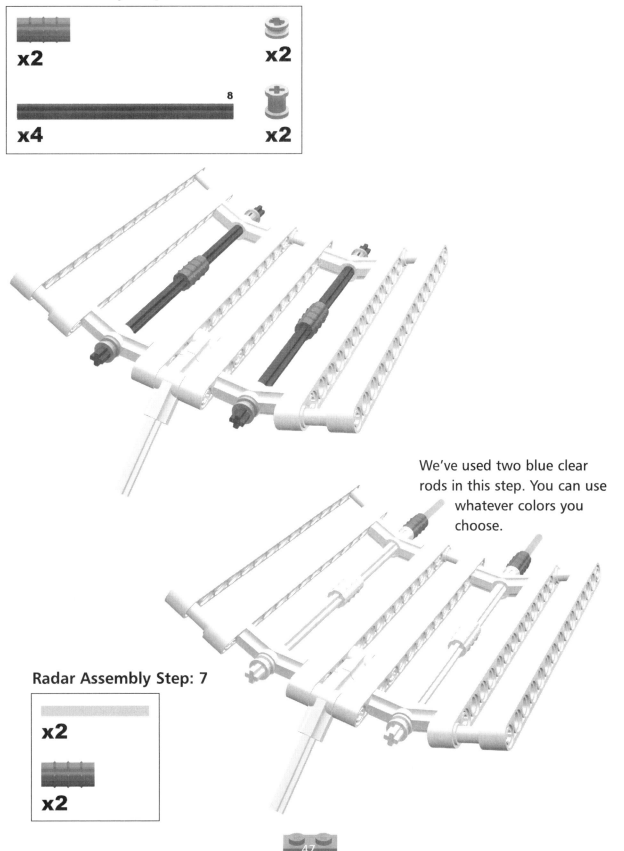

We've used two blue clear rods in this step. You can use whatever colors you choose.

Radar Assembly Step: 7

x2

x2

The Front

Front Assembly Step: Complete

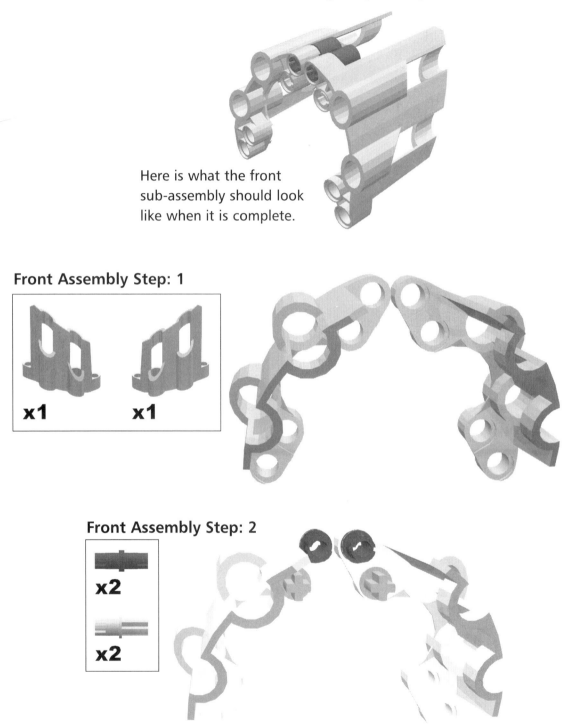

Here is what the front sub-assembly should look like when it is complete.

Front Assembly Step: 1

x1   x1

Front Assembly Step: 2

x2

x2

Front Assembly Step: 3

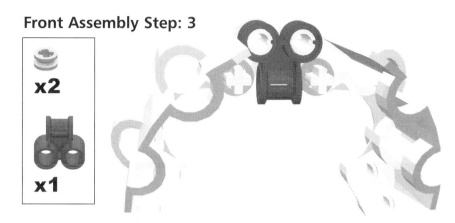

Final Assembly

You are now ready to take all of the sub-assemblies you have built (the right and left lasers, the front, and the radar) and attach them to the main sub-assembly to create the finished Orbital Defense Cannon.

Final Assembly Step: Complete

Here is what the Orbital Defense Cannon should look like when it is complete.

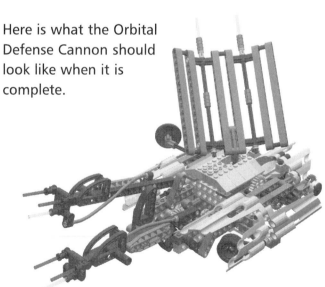

Final Assembly Step: 1

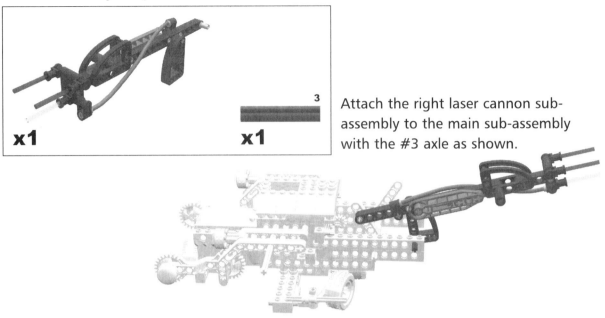

3

x1 x1

Attach the right laser cannon sub-assembly to the main sub-assembly with the #3 axle as shown.

Final Assembly Step: 2

x2

x2

x1

Attach the liftarms as shown. It is important that the 1x3 liftarms point down.

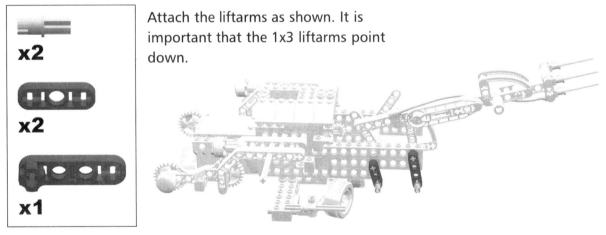

Final Assembly Step: 3

x1

Final Assembly Step: 4

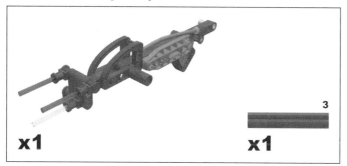

x1   **x1**

Attach the left laser cannon sub-assembly to the main sub-assembly with the #3 axle as shown.

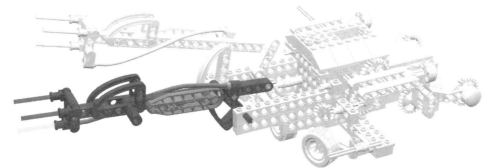

Final Assembly Step: 5

x2

x2

x1

Attach the liftarms as shown. Make sure that the 1x3 liftarms point up.

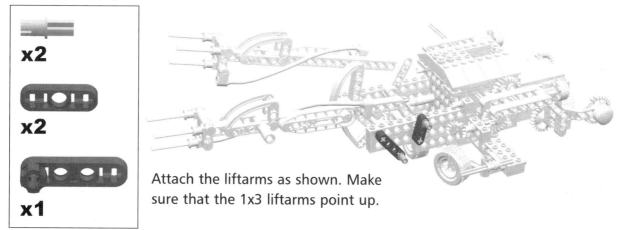

Final Assembly Step: 6

x1

Final Assembly Step: 7

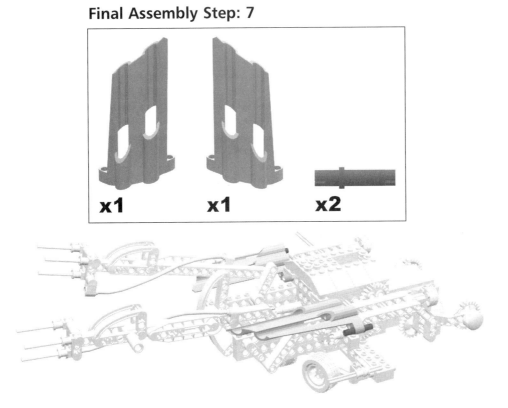

Final Assembly Step: 8

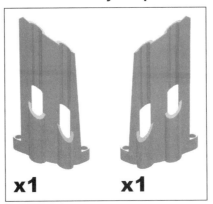

Attach a side panel on each side to the pin you attached in the previous step and to the axle next to the rear wheels as shown.

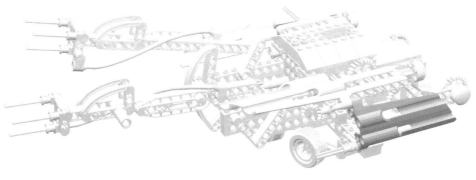

Final Assembly Step: 9

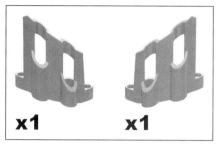

Attach a small panel on each side to the pin you attached in **Final Step: 7** and to the axle next to the wheels as shown.

Final Assembly Step: 10

The radar sub-assembly is attached through the hole in the top of the main sub-assembly. Make sure the axle that makes up the base of the radar sub-assembly fits snugly into the 12t red gear you attached in **Main Step: 26.**

Final Assembly Step: 11

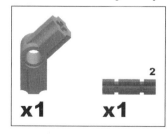

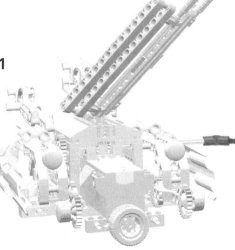

Final Assembly Step: 12

x1   **x1**

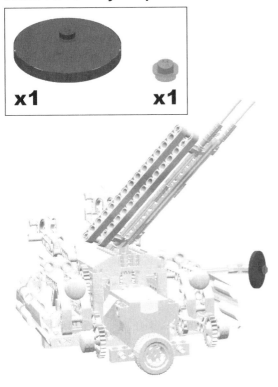

Final Assembly Step: 13

Attach the front sub-assembly to the two pins located on either side of the front wheel as shown.

Final Assembly Step: 14

x2

x2

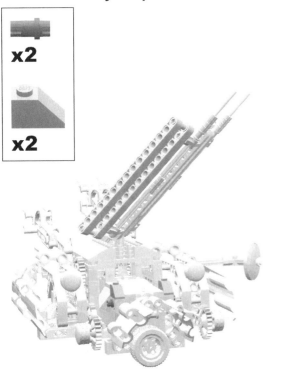

Final Assembly Step: 15

x2

x2

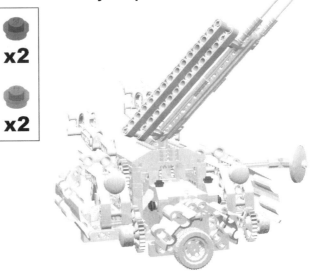

We've used two red and two blue pieces for this step. You can use whatever colors you want. Your Orbital Defense Cannon is complete!

Robot 3

Imperial AT-ST

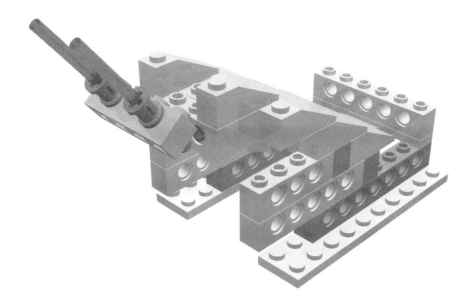

The Imperial All-Terrain Scout Transport, better known as the AT-ST, is a small, agile, armored biped that can carry two Imperial Stormtroopers. Used primarily for scouting and reconnaissance patrols by Imperial troops, the AT-ST is a force to be reckoned with. While not as large or powerful as the AT-AT, its quickness and agility make it a relentless and implacable foe, and its twin laser cannons can make short work of most infantry and light vehicles. AT-STs were used in the battles of Hoth and Endor, as seen in *Episode V: The Empire Strikes Back* and *Episode VI: Return of the Jedi.*

The LEGO Mindstorms version of the AT-ST presented here is an excellent example of a biped robot. Like the other bipeds in this book, the AT-ST walks by shifting its weight from side to side and forward and backward. In some ways, the AT-ST robot is very similar to the Super Battle Droid robot also presented in this book.

Bricks & Chips...

Micro Scout Programs

Since the AT-ST can walk both forward and backward, you can use any of the programs that come with the Micro Scout to power it.

The Body

The body is the heart of the AT-ST. You will construct it first, and then build each of the separate sub-assemblies that will be added to the body to create the finished AT-ST.

Body Step: Complete

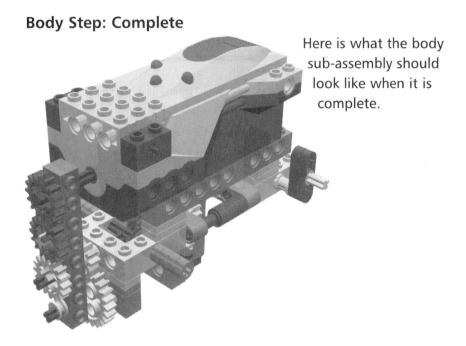

Here is what the body sub-assembly should look like when it is complete.

Body Step: 1

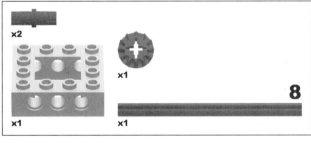

Body Step: 2

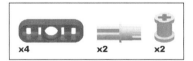

Body Step: 3

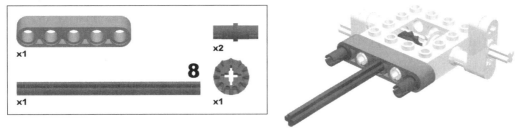

Body Step: 4

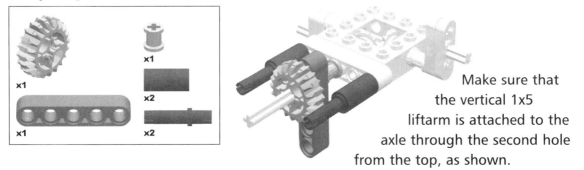

Make sure that the vertical 1x5 liftarm is attached to the axle through the second hole from the top, as shown.

Body Step: 5

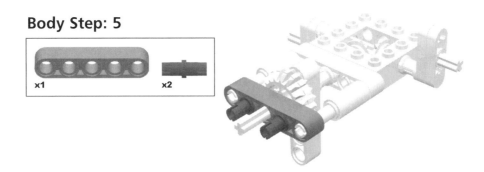

Body Step: 6

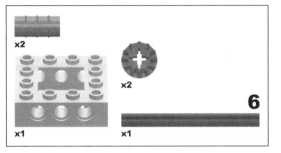

When adding the axle and gears in this step, make sure that the moving parts at the far end are positioned as shown in this image, with one set pointing straight up and the other straight down. It is also important to make sure the gears allow the #6 axle to be oriented as shown.

Body Step: 7

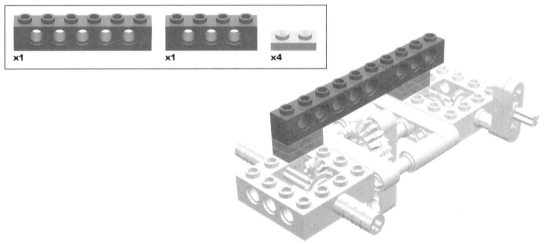

Body Step: 8

x1 x1 x4

Body Step: 9

x1 x2

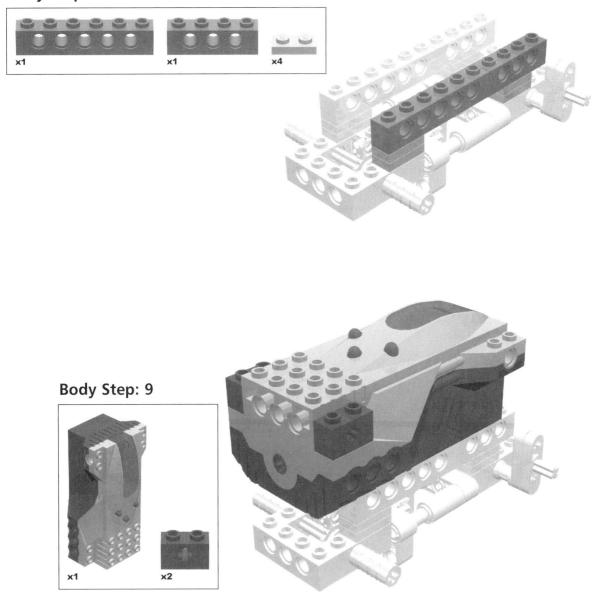

Body Step: 10

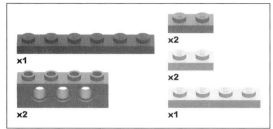

Flip the body over so that the Micro Scout is on the bottom, and add the pieces as shown.

Body Step: 11

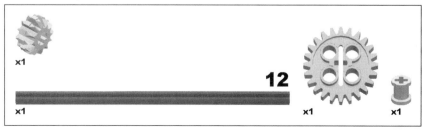

Body Step: 12

Flip the body back over so
that the Micro Scout is again
on the top, and add the
pieces as shown.

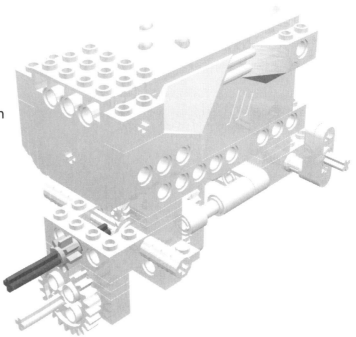

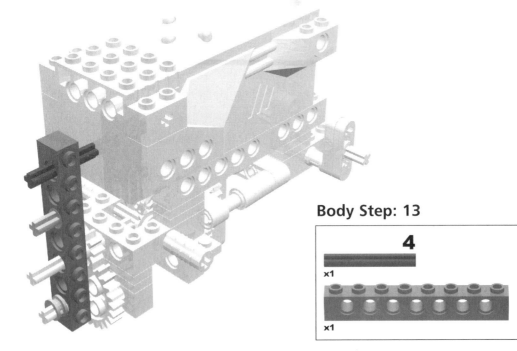

Body Step: 13

Body Step: 14

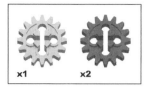

x1   x2

In **Body Step: 14**, you can use any three of the four 16t gears in the DSDK set. We chose to use two red and one gray.

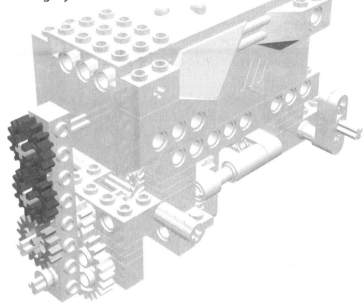

The Right Leg

Right Leg Step: Complete

Here is what the right leg sub-assembly should look like when it is complete.

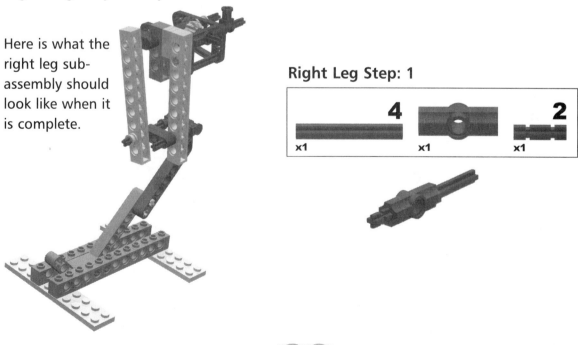

Right Leg Step: 1

4		**2**
x1	x1	x1

Right Leg Step: 2

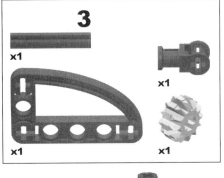

3
x1

x1

x1

x1

Right Leg Step: 3

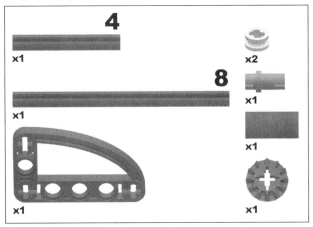

4
x1

8
x1

x1

x1

x2

x1

x1

x1

Make sure
the gears mesh
together as shown.

Right Leg Step: 4

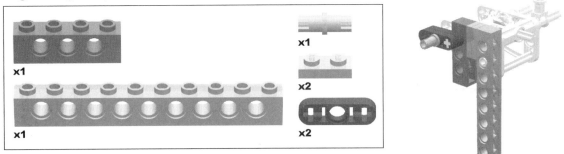

x1

x1

x1

x2

x2

Right Leg Step: 5

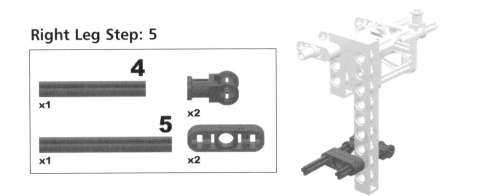

4
x1

5
x1

x2

x2

Right Leg Step: 6

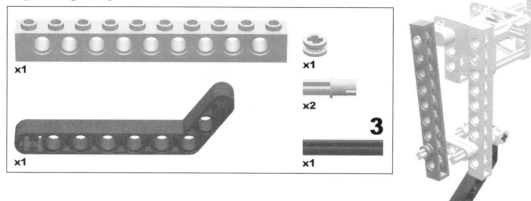

Right Leg Step: 7

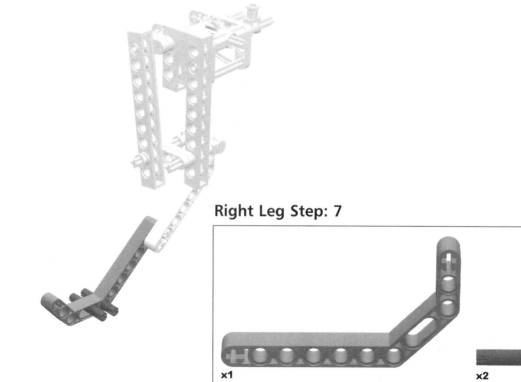

Right Leg Step: 8

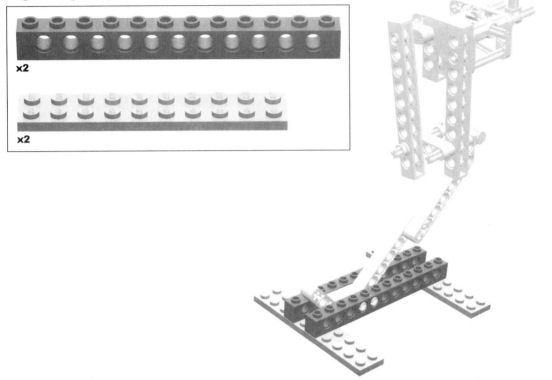

The Left Leg

Left Leg Step: Complete
Here is what the left leg sub-assembly should look like when it is complete.

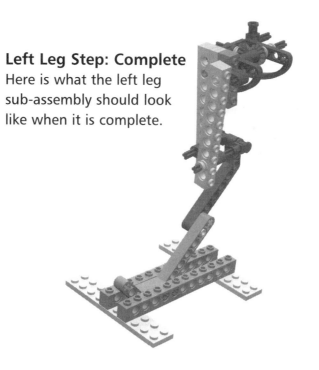

Left Leg Step: 1

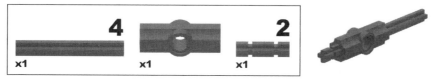

Left Leg Step: 2

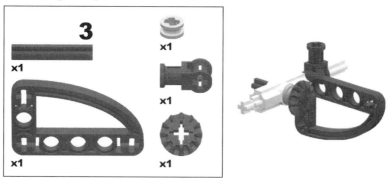

Left Leg Step: 3

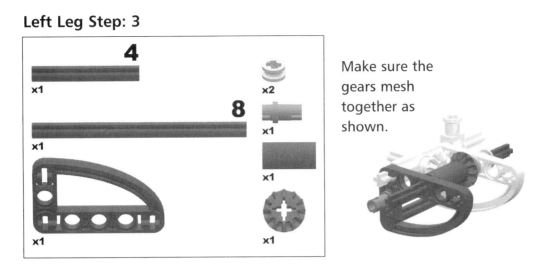

Make sure the gears mesh together as shown.

Left Leg Step: 4

Left Leg Step: 5

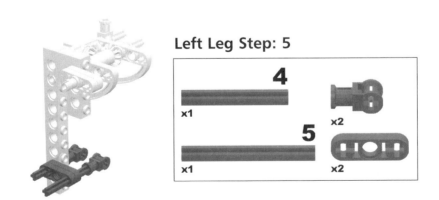

Left Leg Step: 6

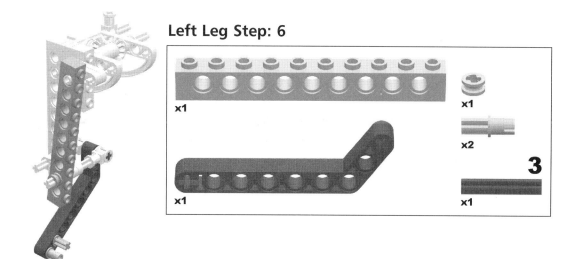

x1

x1

x1

x2

3

x1

Left Leg Step: 7

x1

x2

Left Leg Step: 8

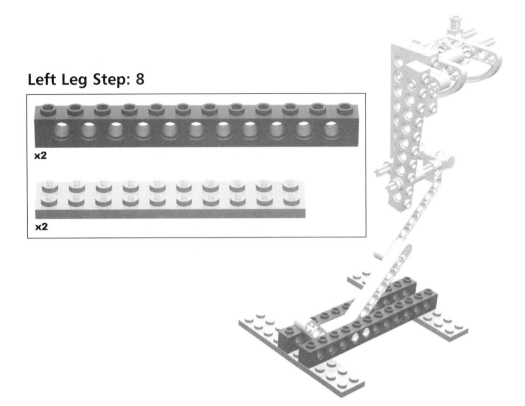

x2

x2

The Face

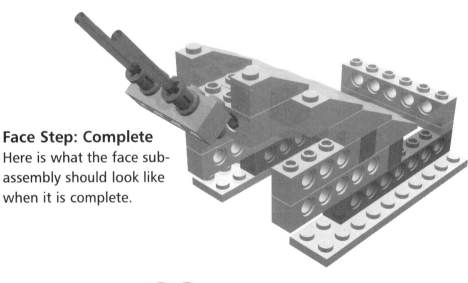

Face Step: Complete
Here is what the face sub-assembly should look like when it is complete.

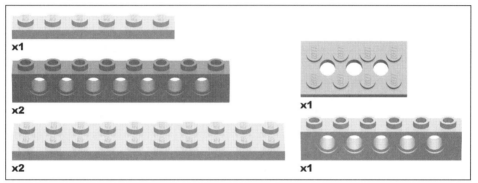

Customizing the Face

You might want to add more decoration to the face, or add weapon pods to the sides of the head to accompany the laser cannons already in place, as on the AT-STs from the *Star Wars* movies. Be careful: Doing so could change the overall balance of the robot. If the AT-ST starts to fall forward on its face, you can compensate for the weight by adjusting the feet farther forward on the legs, or by adding some more bricks to the rear of the AT-ST. Don't be afraid to experiment!

Face Step: 1

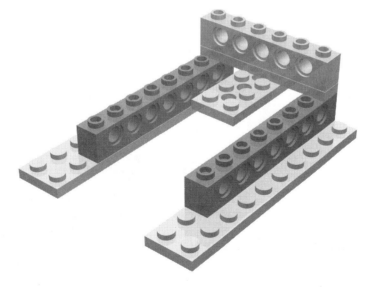

Face Step: 2

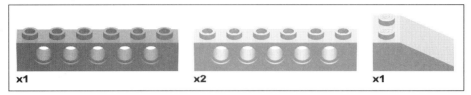

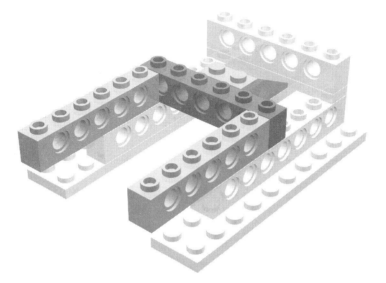

Face Step: 3

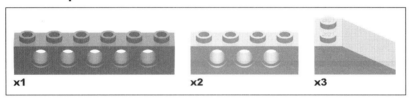

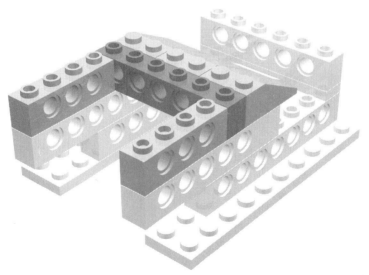

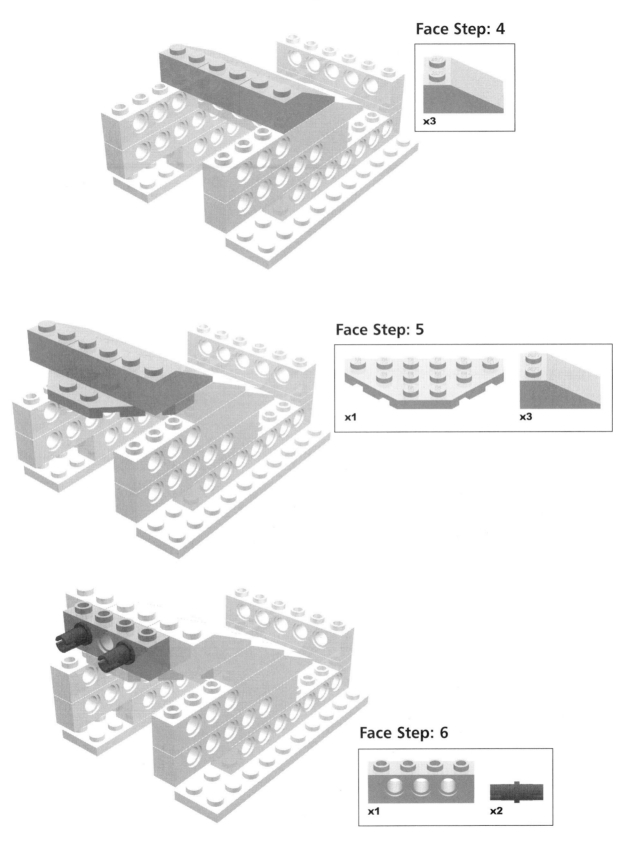

Face Step: 4

x3

Face Step: 5

x1   x3

Face Step: 6

x1   x2

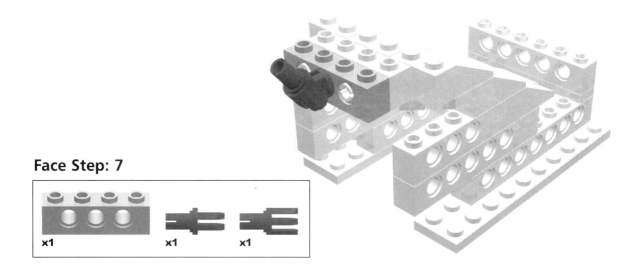

Face Step: 7

x1 x1 x1

Face Step: 8

x2 x2
x1 x2 **R**

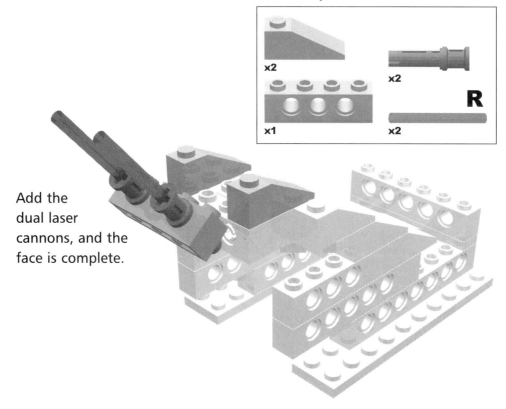

Add the dual laser cannons, and the face is complete.

Final Assembly

You are now ready to take all of the sub-assemblies you have made (the right and left legs and the face), and attach them to the body sub-assembly to create the finished AT-ST.

Final Assembly Step: Complete
Here is what the AT-ST should look like when it is complete.

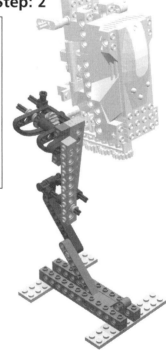

Final Assembly Step: 1
Hold the body with the Micro Scout facing you. Make sure the moving parts on top are aligned as shown.

Final Assembly Step: 2

x1

Attach the right leg to the body. Make sure all the moving parts are aligned as shown.

Final Assembly Step: 3

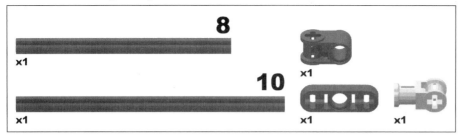

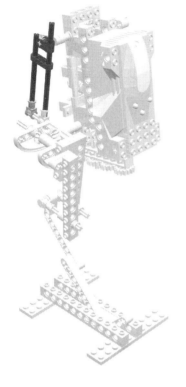

Final Assembly Step: 4

Turn the AT-ST around and attach the left leg to the opposite side. Make sure all the moving parts are aligned as shown.

Final Assembly Step: 5

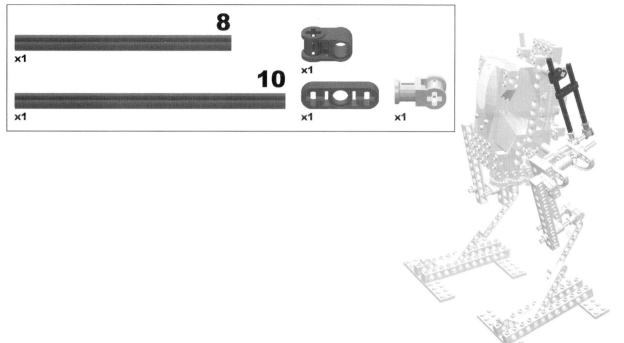

Final Assembly Step: 6

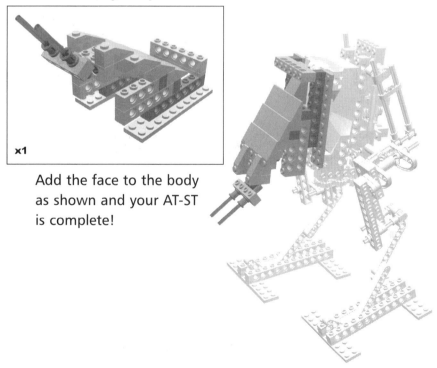

Add the face to the body as shown and your AT-ST is complete!

Robot 4

Go-Rilla

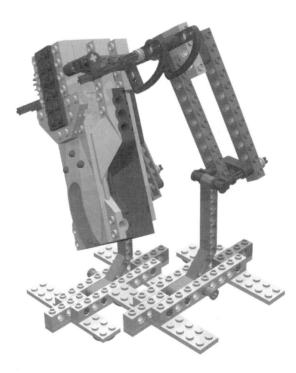

The Go-Rillas are fast biped robots often used by the smugglers and bounty hunters of the galaxy to defend their hideouts. Capable of following only simple commands, a Go-Rilla is not nearly as intelligent as most Droids, but what it lacks in intelligence is more than made up for in speed and brute strength. This robot makes the ideal guardian, capable of running down and trampling with its enormous legs any intruders who enter its field of vision. Stomping around within a territory set by their owners, Go-Rillas will make any intruders think twice before attacking, be they Jedi Masters or a patrol of Imperial Stormtroopers.

The Go-Rilla is the fastest of the three biped robots in this book. Let's take a moment to look at what it is that makes biped robots so unique.

All bipeds have one talent in common: the ability to balance. There are two important concepts behind balancing. The first is the idea of a *footprint* and the other is *center of gravity*. Your robot's footprint is the shape formed by the points of contact with the ground; a biped's footprint is like a box drawn around the supporting "feet." To balance, a biped needs to keep its center of gravity within the boundaries of that footprint. Any weight added outside of the footprint would cause the center of gravity to shift outside the footprint, causing the biped to fall over–unless it can be balanced out by a weight outside the opposite side of the footprint that shifts the center of gravity back inside its boundaries.

The center of gravity has three dimensions that affect it: *side-to-side, front-to-back* and *top-to-bottom*. Super Battle Droid, AT-ST, and Go-Rilla each have similar side-to-side and front-to-back dimensions that affect their centers of gravity. However, a Go-Rilla's top-to-bottom dimension is much lower to the ground than the other two bipeds, which makes it more stable. Also, a Go-Rilla's top-to-bottom center of gravity is very near its hips (where it leans side to side), which allows it to shift its weight side-to-side much faster than the Super Battle Droid or AT-ST. These factors allow it to take many more steps per minute than the other two bipeds.

Bricks & Chips...

Micro Scout Programs

Go-Rilla can only walk forward, so it should be used with Micro Scout program number 1.

The Drive

The drive sub-assembly is the most important sub-assembly of the Go-Rilla. The drive sub-assembly includes both the body of the Go-Rilla and the large hips. Later, you will attach to the hips the huge legs that are so characteristic of the Go-Rilla and LEGO Mindstorms bipeds in general.

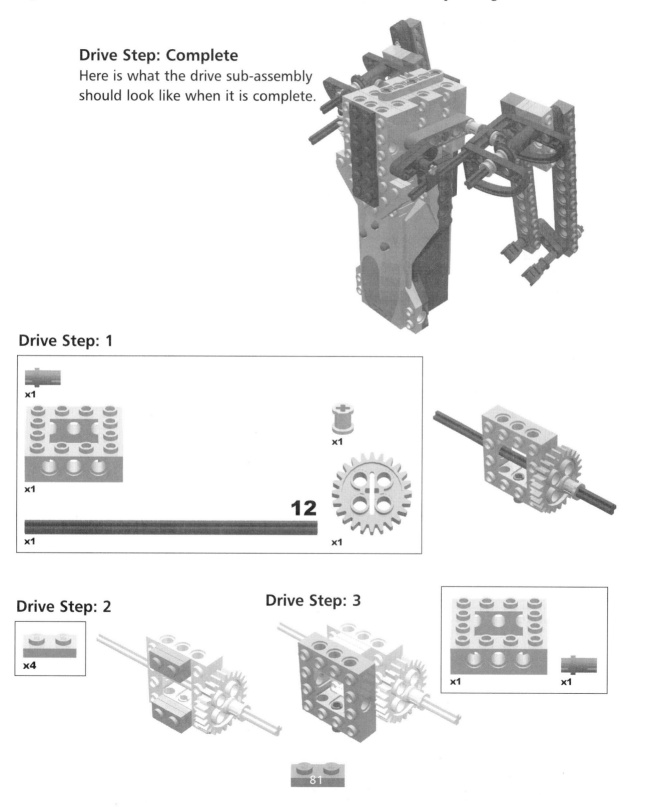

Drive Step: Complete

Here is what the drive sub-assembly should look like when it is complete.

Drive Step: 1

x1

x1

x1

x1

12

x1

Drive Step: 2

x4

Drive Step: 3

x1

x1

Drive Step: 4

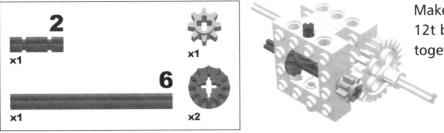

Make sure that the two 12t beveled gears mesh together as shown.

Drive Step: 5

Drive Step: 6

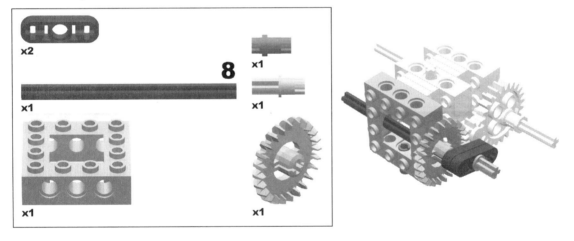

Drive Step: 7

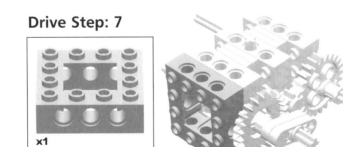

Drive Step: 8

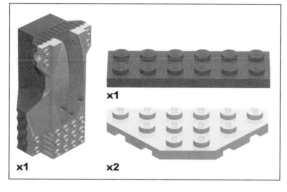

Attach the Micro Scout to the drive sub-assembly as shown.

Drive Step: 9

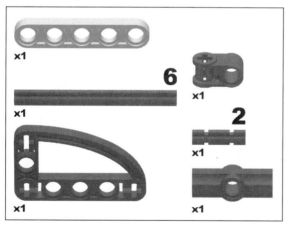

Turn the drive sub-assembly over so the Micro Scout is upside down and begin construction on the right hip. Because the drive sub-assembly is upside down, the right hip will be on the left side of the drive sub-assembly when the Micros Scout is facing you.

Drive Step: 10

x2

x3

8

x1

x1

Make sure all the moving parts
are aligned as shown.

Drive Step: 11

x1

Drive Step: 12

x2

x1

x1

x1

Rotate the drive sub-assembly and
begin construction of the left hip so
that it mirrors the right hip.

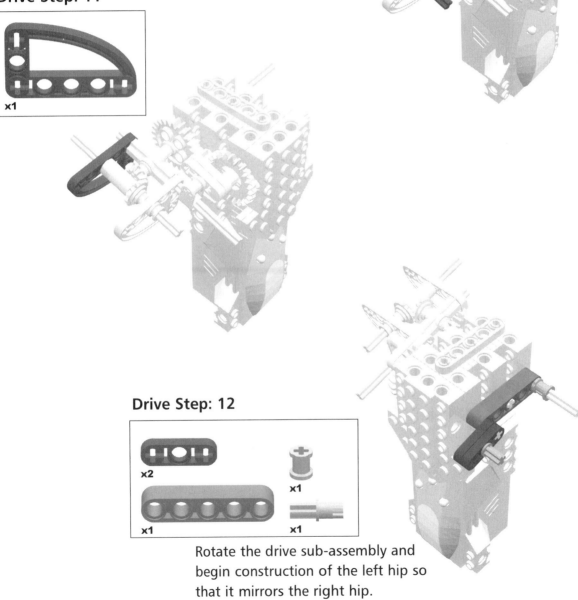

Drive Step: 13

6 x1

2 x1

x1

x1

x1

Drive Step: 14

x2

x3

8 x1

Drive Step: 15

x1

x1

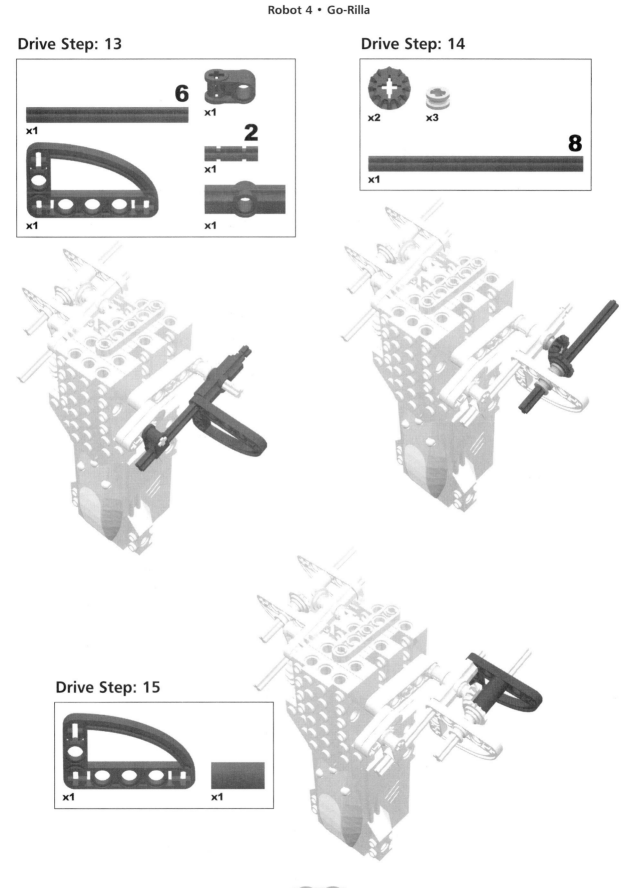

Drive Step: 16

Rotate the drive sub-assembly
and add these pieces as
shown (this is the back
of the Go-Rilla).

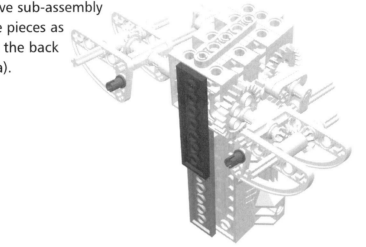

Drive Step: 17

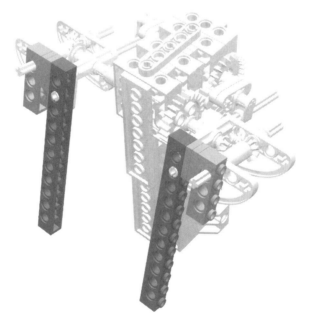

Drive Step: 18

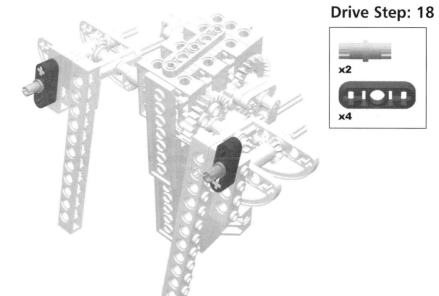

x2

x4

Drive Step: 19

3
x1

5
x1

x1

x1

x1

x2

x2

Take a look at the smaller image as well; it shows how all of these pieces fit together from the front view of the drive sub-assembly.

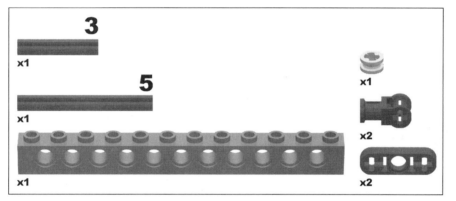

Drive Step: 20

Again, take a look at the smaller image; it shows how all of these pieces fit together from the front view of the drive sub-assembly.

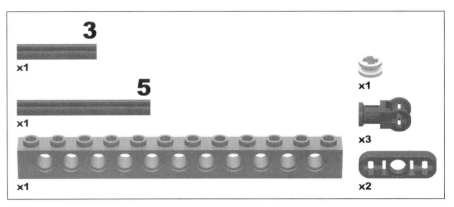

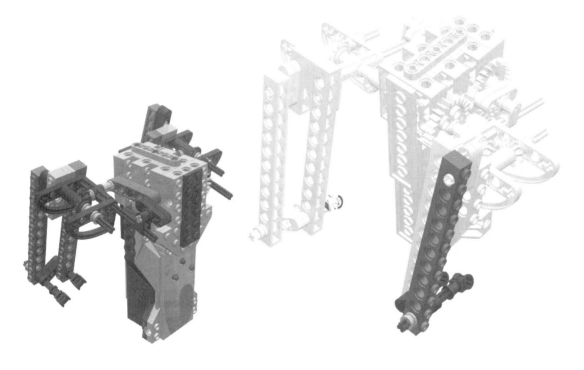

Inventing...

Customizing the Go-Rilla

The Go-Rilla's center of gravity plays a very important role in allowing him to move so quickly—however, this also makes it one of the most difficult Droids in this book to customize. Any additions to one side of the Go-Rilla will have to be balanced out on the opposite side. For example, if you add some facial features to the front of the Go-Rilla, you will need to add something to its back as well to compensate for the shift in weight.

The Right Leg

Right Leg Step: Complete

Here is what the right leg sub-
assembly should look like
when it is complete.

Right Leg Step: 1

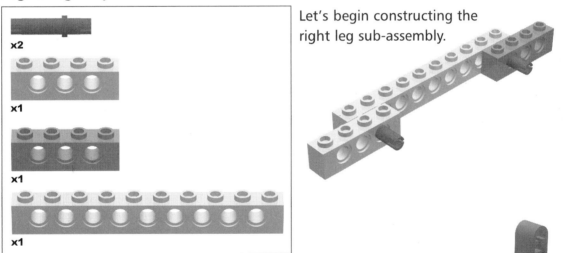

Let's begin constructing the
right leg sub-assembly.

Right Leg Step: 2

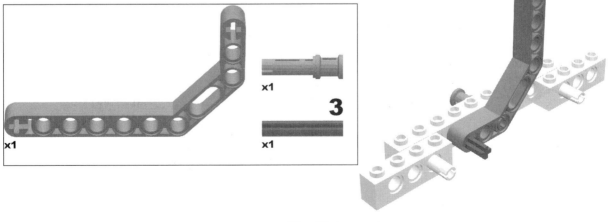

Right Leg Step: 3

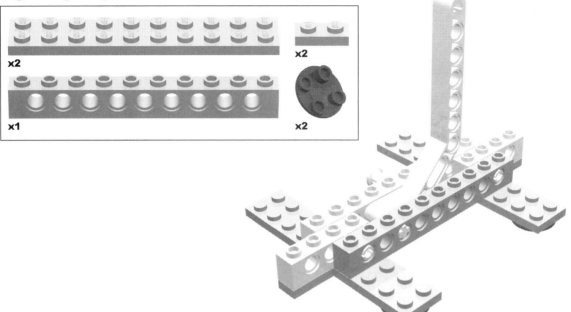

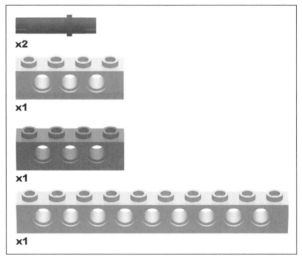

The Left Leg

Left Leg Step: Complete

Here is what the left leg sub-assembly should look like when it is complete.

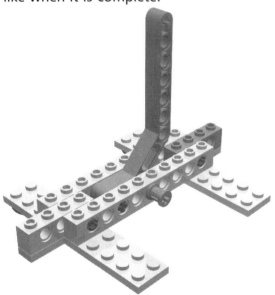

Left Leg Step: 1

Let's begin constructing the left leg sub-assembly.

Left Leg Step: 2

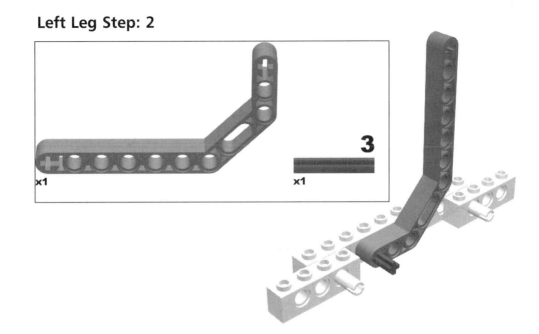

Left Leg Step: 3

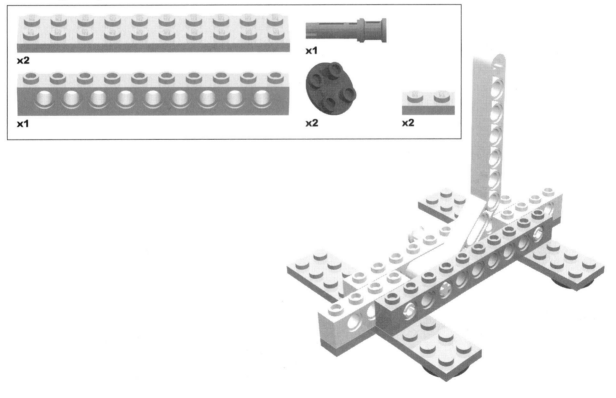

Final Assembly

You are now ready to take all of the sub-assemblies you have made (the right and left leg sub-assemblies), and attach them to the drive sub-assembly.

Final Assembly Step: Complete

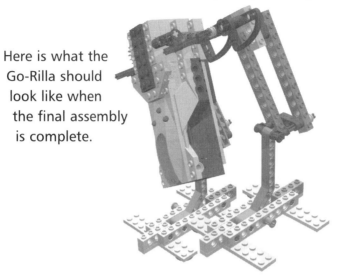

Here is what the Go-Rilla should look like when the final assembly is complete.

Final Assembly Step: 1

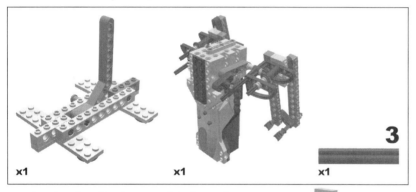

Begin the final assembly by using a #3 axle to attach the right leg sub-assembly to the drive sub-assembly as shown.

Final Assembly Step: 2

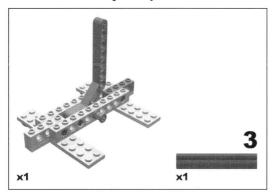

x1   x1   **3**

Use a #3 axle to attach the left leg sub-assembly to the drive sub-assembly as shown, and the Go-Rilla is complete!

Robot 5

X-Stormer

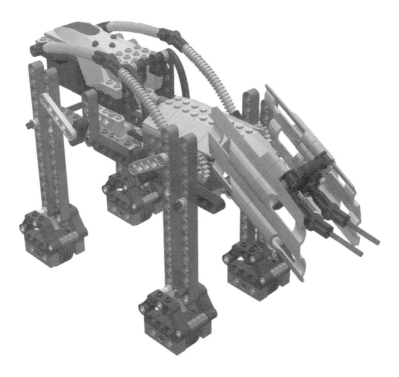

The X-Stormer, a four-legged transport and combat vehicle, was a prototype of what would eventually become the AT-AT. Imperial engineers worked for many years to create the X-Stormers, but once the newer AT-ATs were introduced to the Imperial Army, the X-Stormers were quickly phased out. However, many of them ended up on the black market, where they were picked up by a variety of interested buyers. Rumor has it that the Hutts own several of these surplus X-Stormers, although what they intend to do with them is unknown.

The X-Stormer needs two rubber bands to make its head bounce up and down as it jumps. We recommend small rubber bands like the ones orthodontists use in dental work if you can find them.

Bricks & Chips...

Micro Scout Programs

You can use any of the Micro Scout's programs with the X-Stormer.

The Body

Body Step: Complete

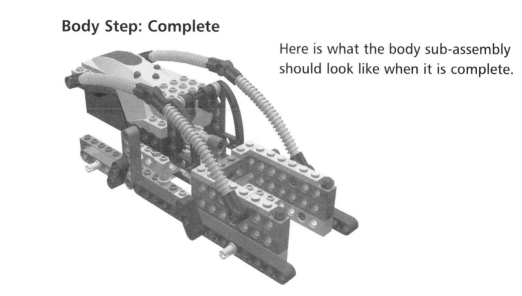

Here is what the body sub-assembly should look like when it is complete.

Body Step: 1

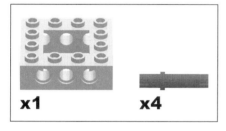

x1   **x4**

Body Step: 2

x1   **x1**

8

x1   **x1**

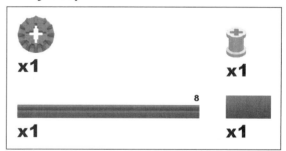

Body Step: 3

x1   **x1**

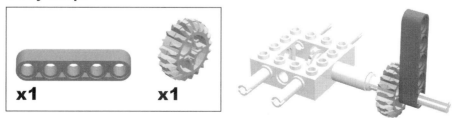

Body Step: 4

x1   x1   x4

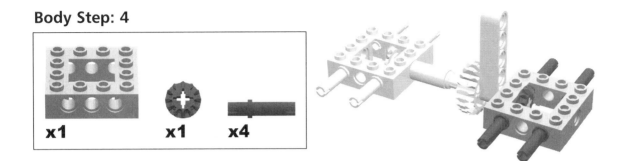

Body Step: 5

x2

x2

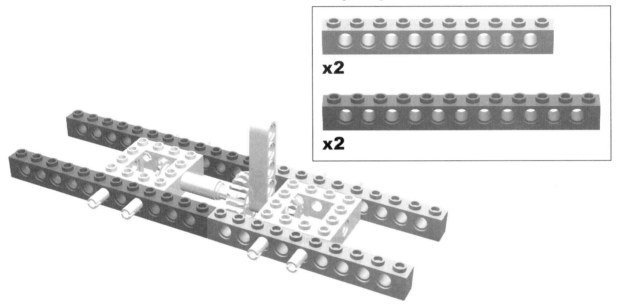

Body Step: 6

x10

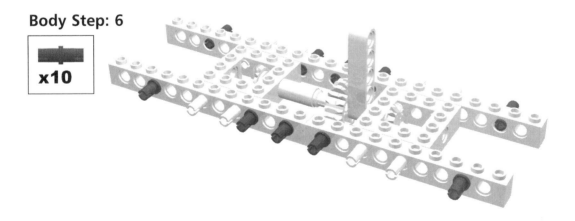

Body Step: 7

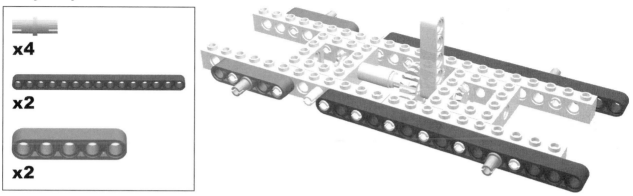

x4

x2

x2

Body Step: 8

x1   x4

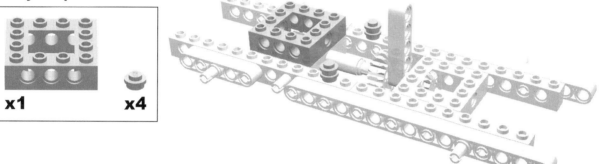

Body Step: 9

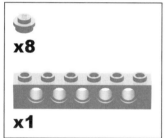

x8

x1

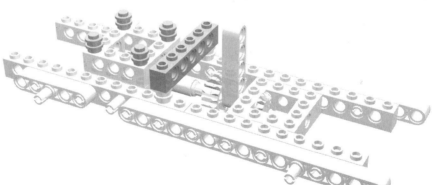

Body Step: 10

x2

x4

x1

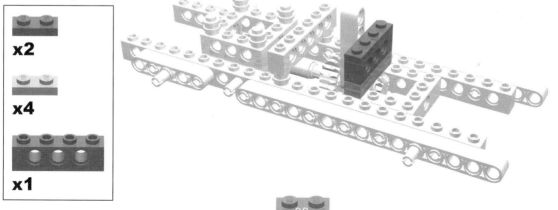

Body Step: 11

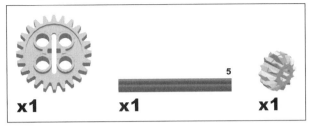

x1   x1   x1

Insert the #5 axle through the middle hole of th 1x3 brick and the gears as shown.

Body Step: 12

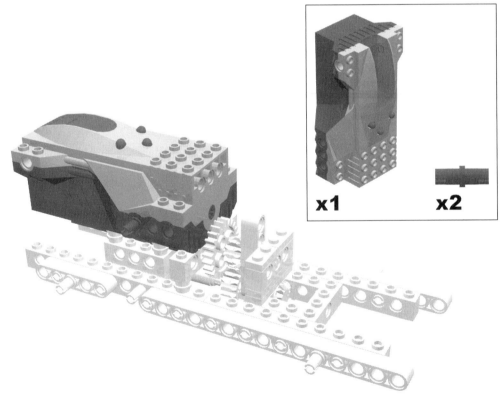

x1   x2

Body Step: 13

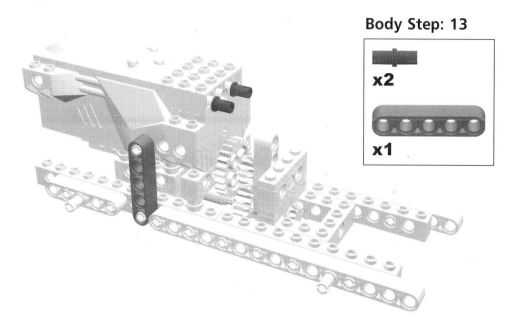

x2

x1

Body Step: 14

x1

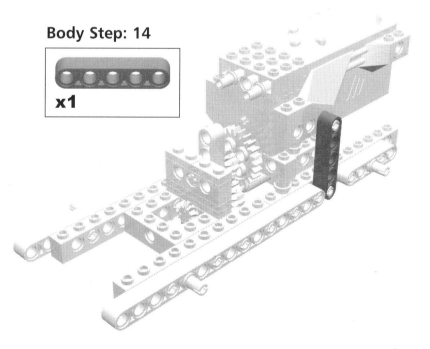

Body Step: 15

x2

x2

5

x1

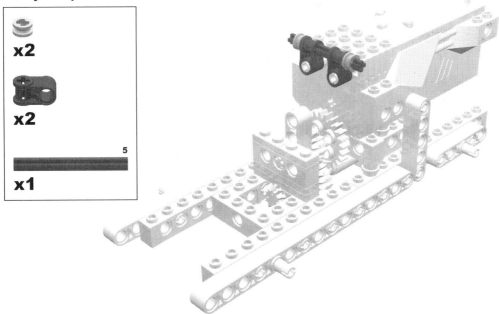

Body Step: 16

x1

x1

4

x1

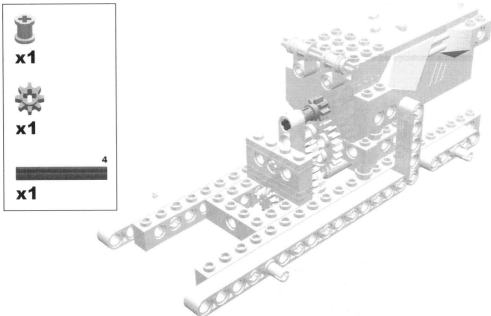

Body Step: 17

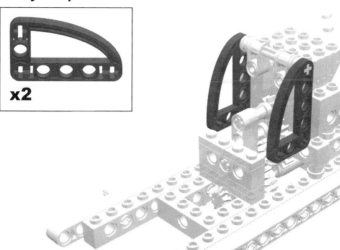

x2

Body Step: 18

x4

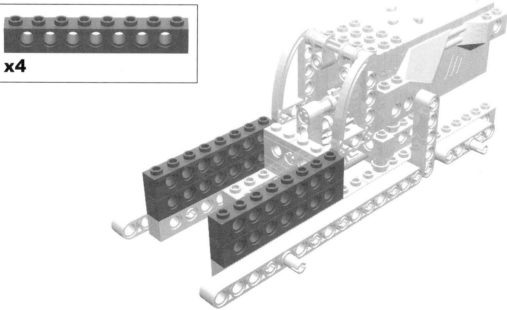

Body Step: 19

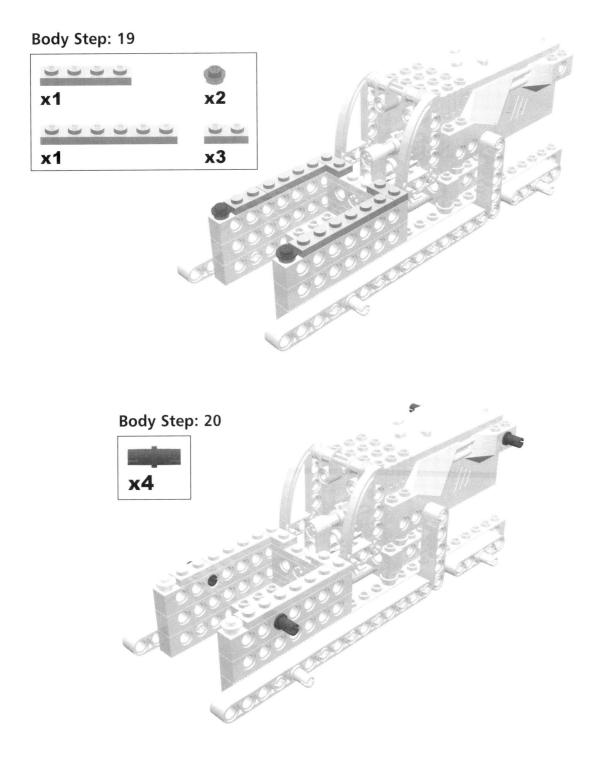

Body Step: 20

Body Step: 21

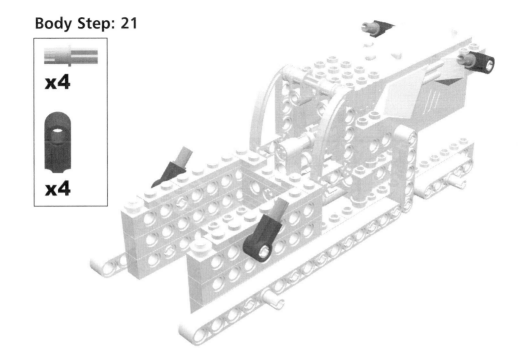

Body Step: 22

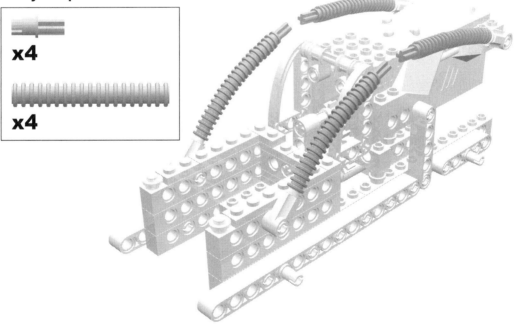

Body Step: 23

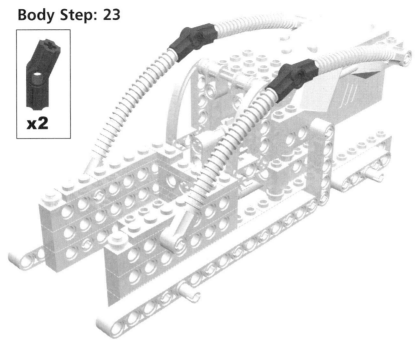

Body Step: 24

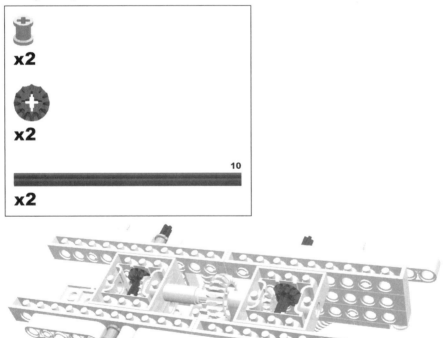

Body Step: 25

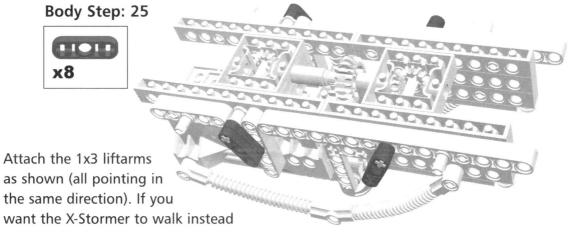

x8

Attach the 1x3 liftarms
as shown (all pointing in
the same direction). If you
want the X-Stormer to walk instead
of jump, put the 1x3 liftarms on one side pointing in the opposite direction from
those on the other side (i.e. pointing up on one side and down on the other).

The Front Laser

Front Laser Step: Complete

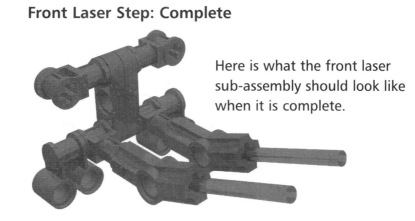

Here is what the front laser
sub-assembly should look like
when it is complete.

Front Laser Step: 1

x2

3

x1

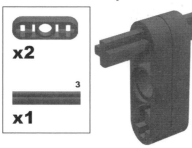

Front Laser Step: 2

x4

5

x1

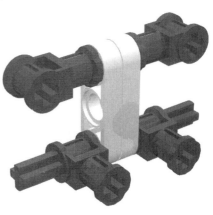

Front Laser Step: 3

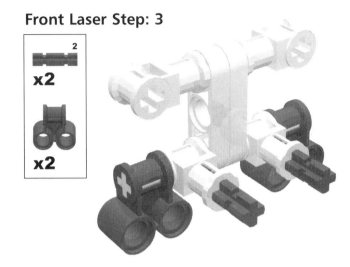

Front Laser Step: 4

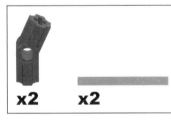

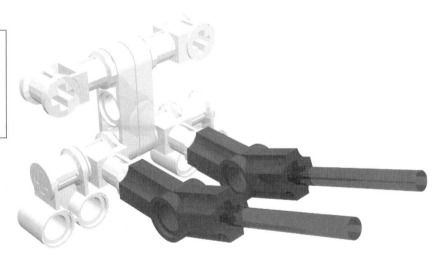

The Front

The front sub-assembly makes up the X-Stormer's head. We will attach the front laser sub-assembly to the front sub-assembly to complete it.

Front Step: Complete

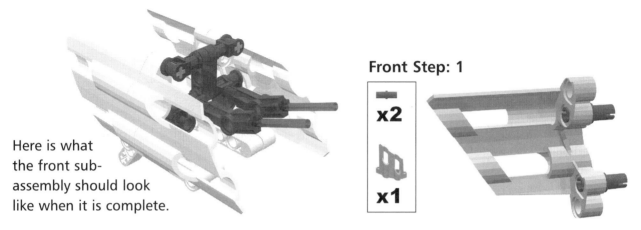

Here is what the front sub-assembly should look like when it is complete.

Front Step: 1

Front Step: 2

Front Step: 3

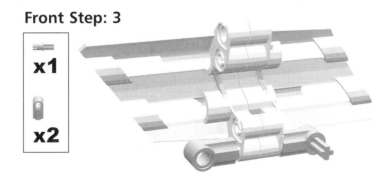

Front Step: 4

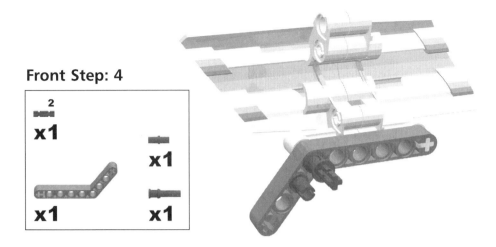

Front Step: 5

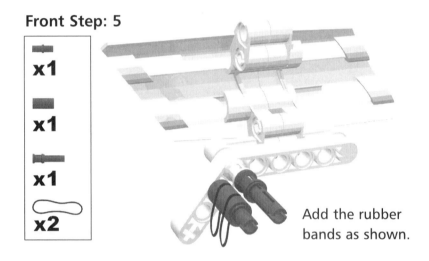

Add the rubber bands as shown.

Front Step: 6

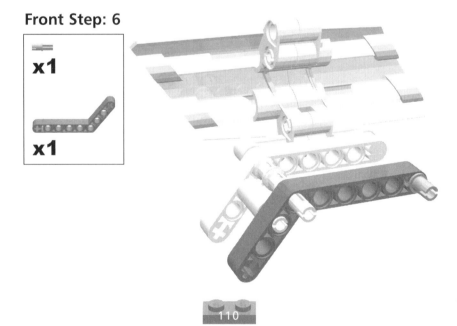

Front Step: 7

x2

x2   x2

x2

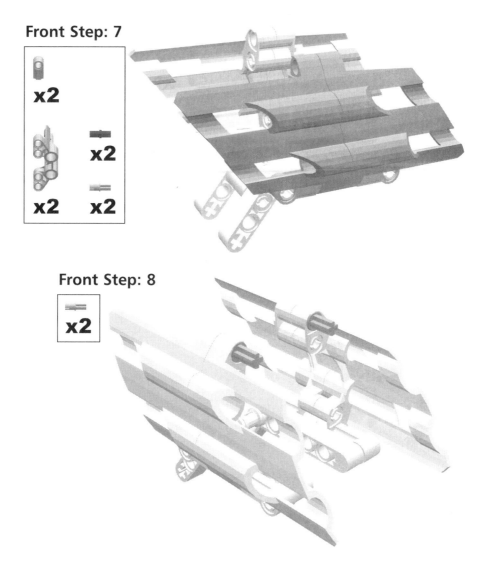

Front Step: 8

x2

Front Step: 9

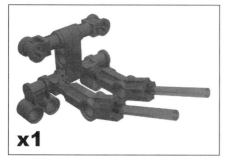

x1

Attach the
front laser sub-
assembly as shown.

The Top

Top Step: Complete

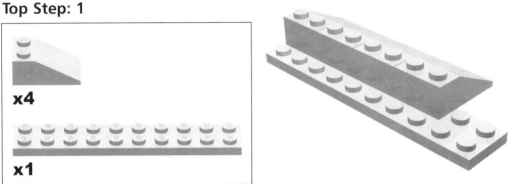

Here is what the top sub-assembly should look like when it is complete.

Top Step: 1

x4

x1

Top Step: 2

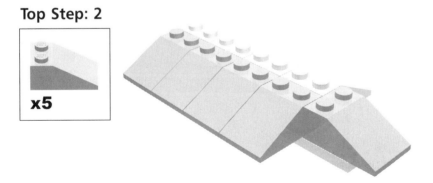

x5

The Right Leg

Right Leg Step: Complete

Here is what the right leg sub-assembly should look like when it is complete. You need to make two right legs.

Right Leg Step: 1

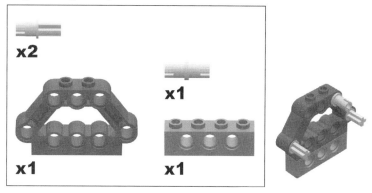

x2

x1

x1

x1

Right Leg Step: 2

x2

x2

Right Leg Step: 3

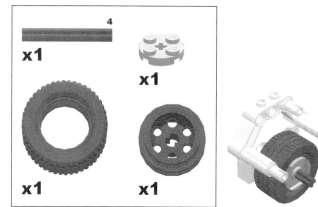

x1 **4**

x1

x1

x1

Right Leg Step: 4

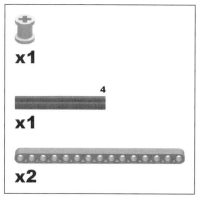

x1

x1 **4**

x2

Right Leg Step: 5

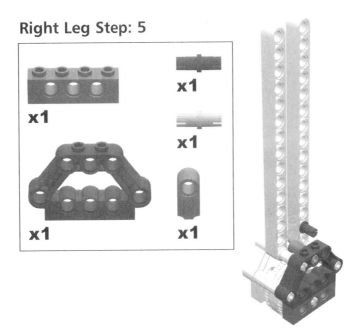

The Left Leg

Left Leg Step: Complete

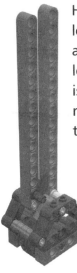

Here is what the left leg sub-assembly should look like when it is complete. You need to make two left legs.

Left Leg Step: 1

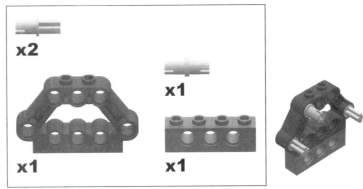

Left Leg Step: 3

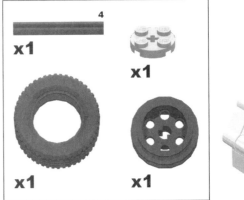

Left Leg Step: 2

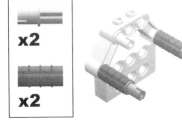

Left Leg Step: 4

x1

4

x1

x2

Left Leg Step: 5

x1

x1

x1

x1

x1

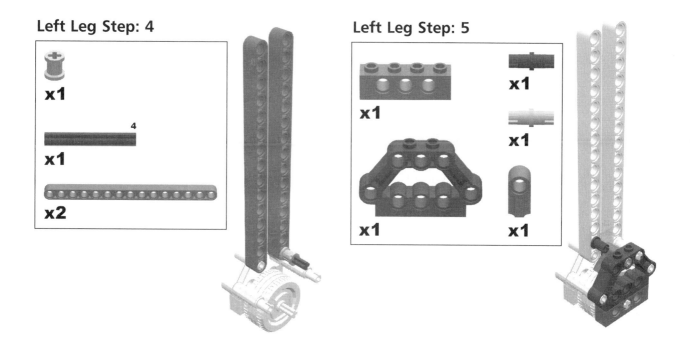

Final Assembly

You are now ready to take all of the sub-assemblies you have built (the front, the top, and the right and left legs) and attach them to the body sub-assembly to make the finished X-Stormer.

Final Assembly Step: Complete

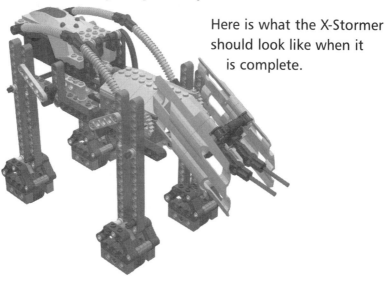

Here is what the X-Stormer should look like when it is complete.

Final Assembly Step: 1

Take the body sub-assembly and turn it on its back as shown.

Final Assembly Step: 2

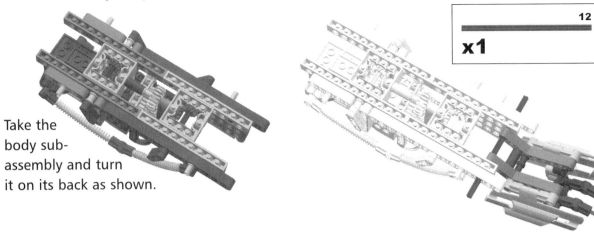

Final Assembly Step: 3

Attach the rubber bands as shown.

Final Assembly Step: 4

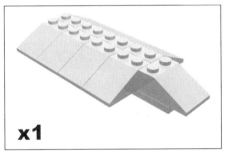

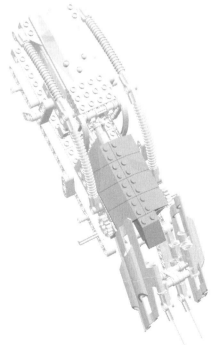

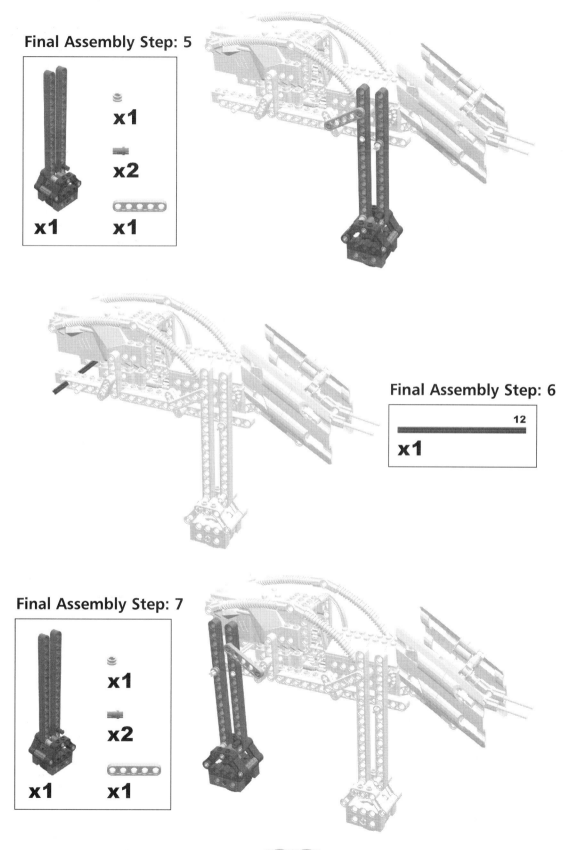

Final Assembly Step: 5

x1

x2

x1 x1

Final Assembly Step: 6

12

x1

Final Assembly Step: 7

x1

x2

x1 x1

Final Assembly Step: 8

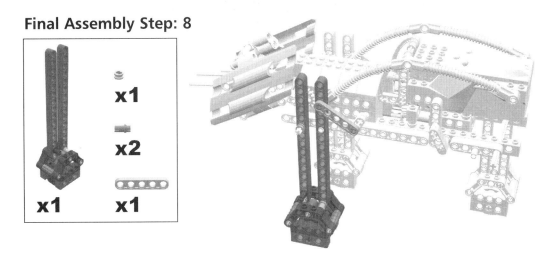

Final Assembly Step: 9

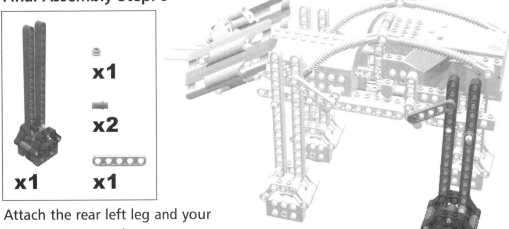

Attach the rear left leg and your X-Stormer is complete!

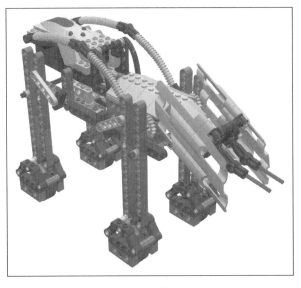

Robot 6

Droid Transporter

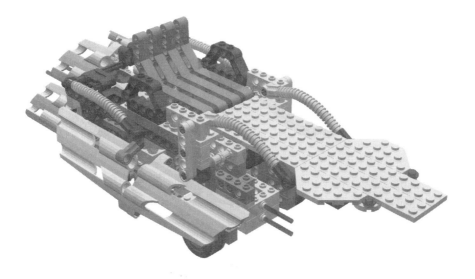

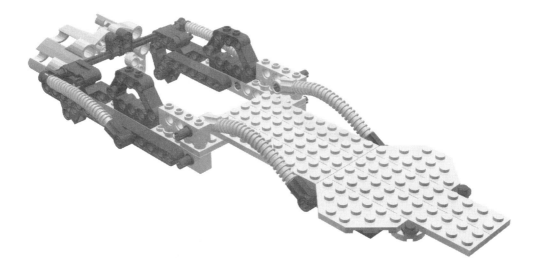

The Droid Transporter is a common vehicle found on most planets in the galaxy. With its heavy wheels and long loading ramp, the Droid Transporter can carry up to a dozen Droids over rough terrain and deliver them where they are needed. Since the Trade Federation and other groups often need to deliver Battle Droids to dangerous areas, some models come equipped with shielding and laser cannons.

The Droid Transporter has two built-in switches that control where the Micro Scout will supply power during a given action:

- When both switches are set in the down position, the Droid Transporter will drive forward and backward; when the motor is in reverse, the ramp will be lowered.

- When the right switch is set in the down position and the left switch is set in the up position, the Droid Transporter will drive forward and backward, but the ramp will not be lowered.

- When the right switch is set in the up position and the left switch is set in the down position, the Droid Transporter will raise and lower the ramp (without moving forward or backward).

You will build these switches in Body Steps 25 and 26.

Bricks & Chips...

Micro Scout Programs

The Droid Transporter works best with Micro Scout programs #2 and #3 (program #3 requires a light source).

The Body

The body sub-assembly of the Droid Transporter takes up the majority of the building process.

Body Step: Complete

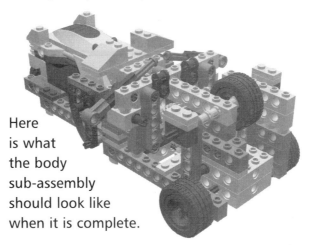

Here
is what
the body
sub-assembly
should look like
when it is complete.

Body Step: 1

Body Step: 2

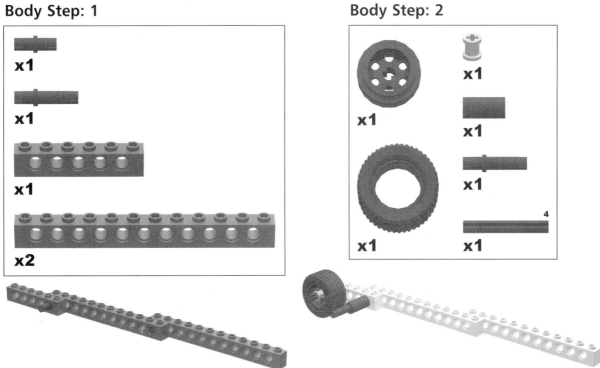

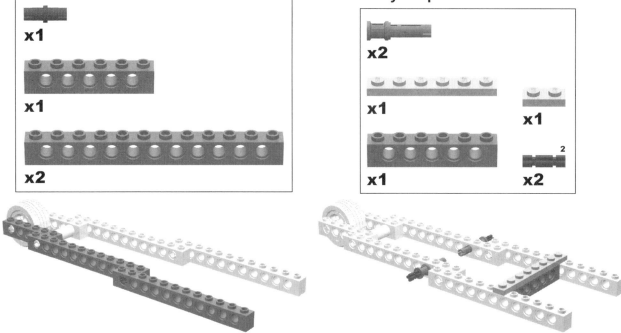

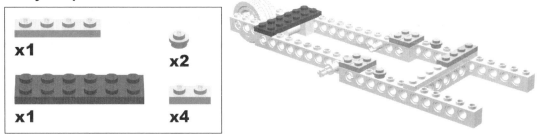

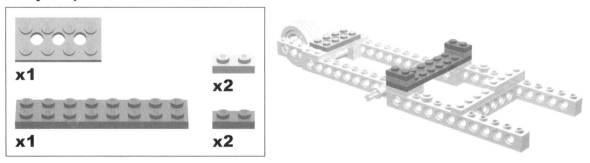

Body Step: 3

x1

x1

x2

Body Step: 4

x2

x1

x1

x1

x1

2

x2

Body Step: 5

x1

x2

x1

x4

Body Step: 6

x1

x2

x1

x2

Body Step: 7

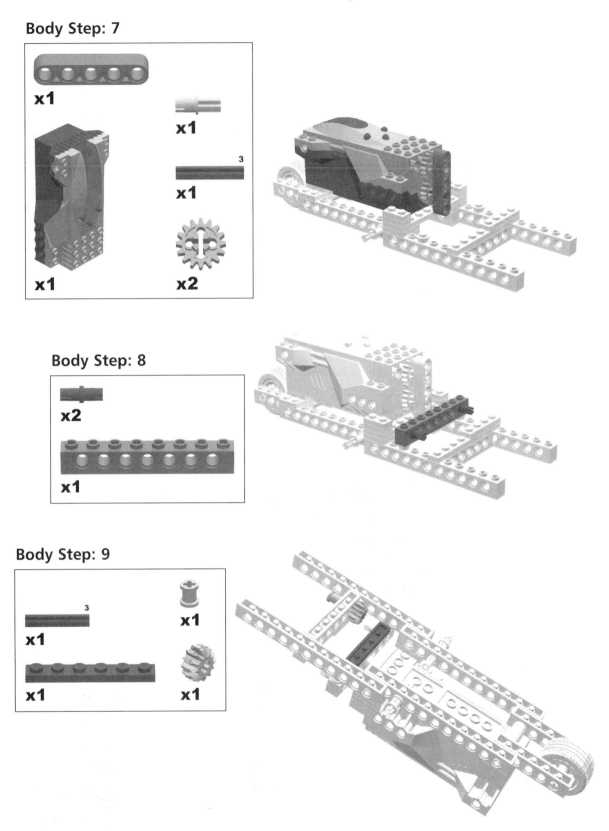

Body Step: 8

Body Step: 9

Body Step: 10

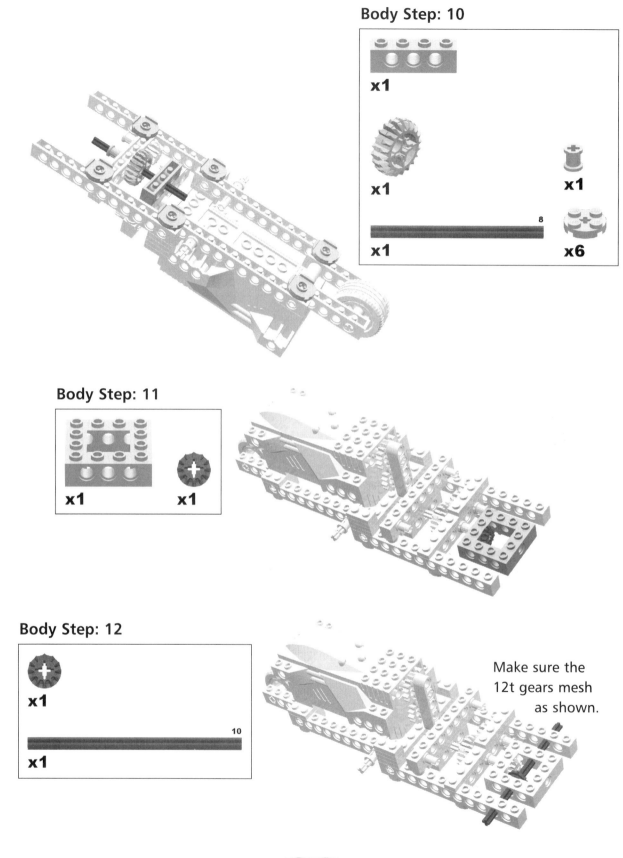

x1

x1

x1

8
x1

x1

x6

Body Step: 11

x1

x1

Body Step: 12

x1

10
x1

Make sure the
12t gears mesh
as shown.

Body Step: 13

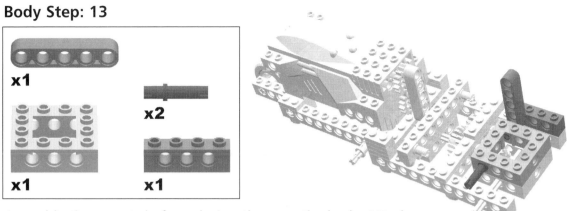

Assemble these parts before placing them on the body. Attach both pins as shown. The second pin attaches the 1x5 liftarm and continues through to the 1x4 brick.

Body Step: 14

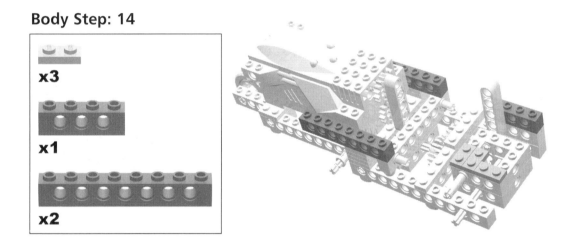

Body Step: 15

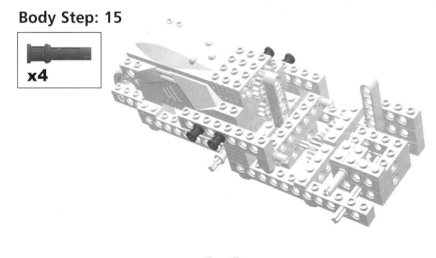

Body Step: 16

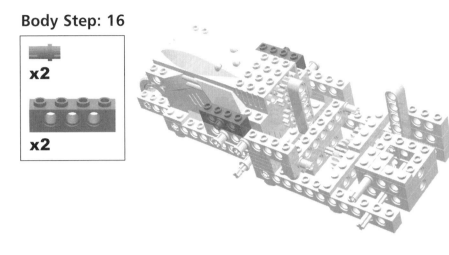

Body Step: 17

Body Step: 18

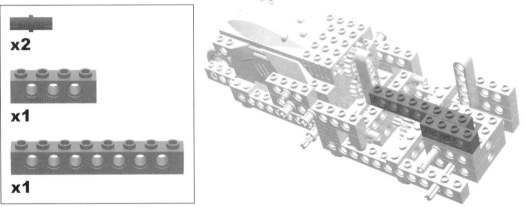

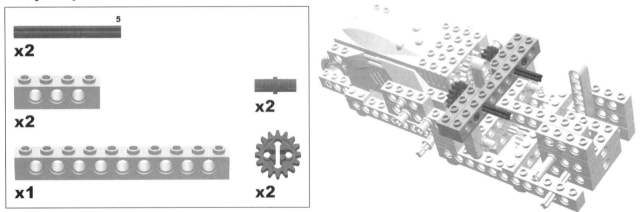

Assemble these parts before placing them on the body. Make sure that the 1x5 liftarm next to the gears can move freely from side to side.

Body Step: 19

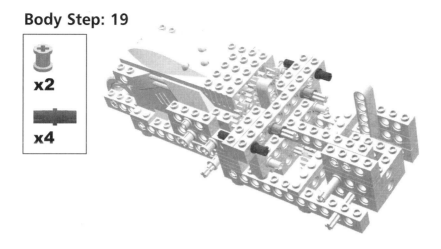

Body Step: 20

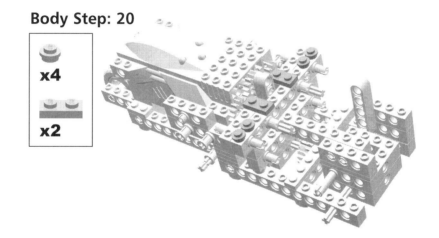

Body Step: 21

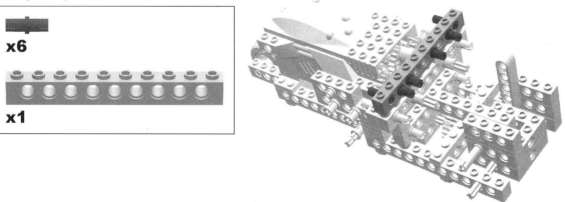

Body Step: 22

x2

x2

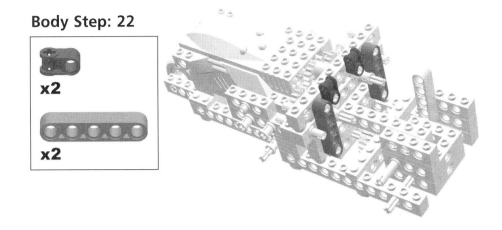

Body Step: 23

x1 **x1** **x1**

Assemble these parts before placing them on the body.

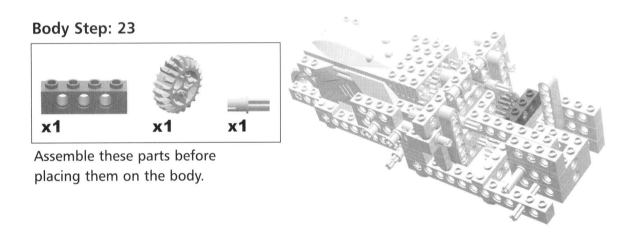

Body Step: 24

x6

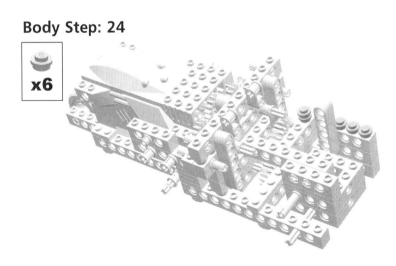

Body Step: 25

x2   **x2**   **x2**

You are now making the two switches that allow the Droid Transporter go forward and backward, and to raise and lower the ramp. Setting these in the up or down position will decide which actions occur (see the beginning of the chapter for more details).

Body Step: 26

x1

x2

x1

Body Step: 27

x1

x4

x1

x1

Assemble these parts before placing them on the body.

Body Step: 28

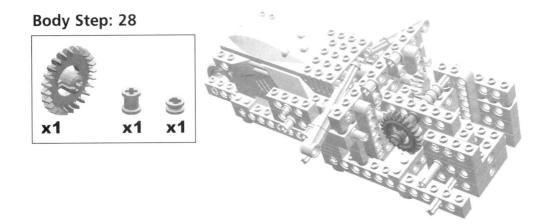

x1   x1   x1

Body Step: 29

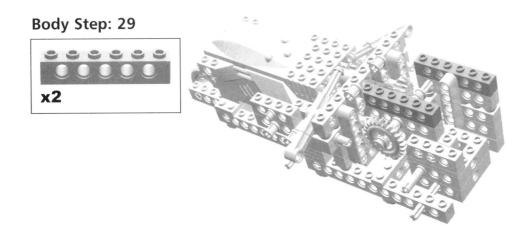

x2

Body Step: 30

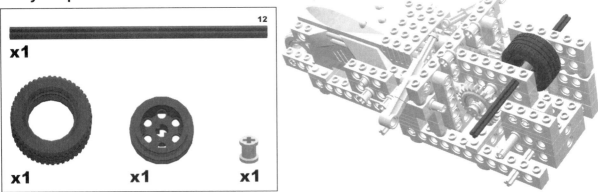

12

x1

x1   x1   x1

Body Step: 31

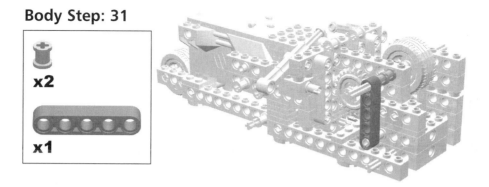

Body Step: 32

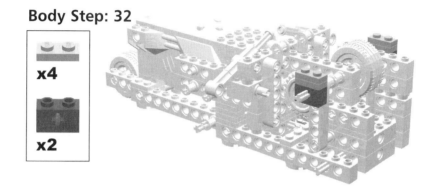

Body Step: 33

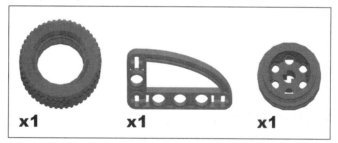

Body Step: 34

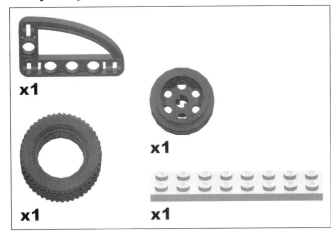

x1

x1

x1

x1

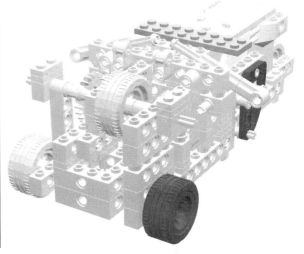

Body Step: 35

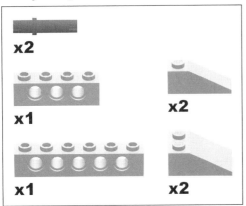

x2

x1

x1

x2

x2

Body Step: 36

x6

x2

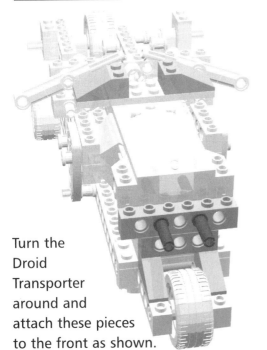

Turn the
Droid
Transporter
around and
attach these pieces
to the front as shown.

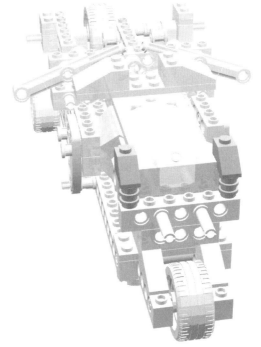

The Platform

Platform Step: Complete

Here is what the platform sub-assembly should look like when it is complete.

Platform Step: 1

x1

6

x2

x2

Platform Step: 2

x1   x2   x1

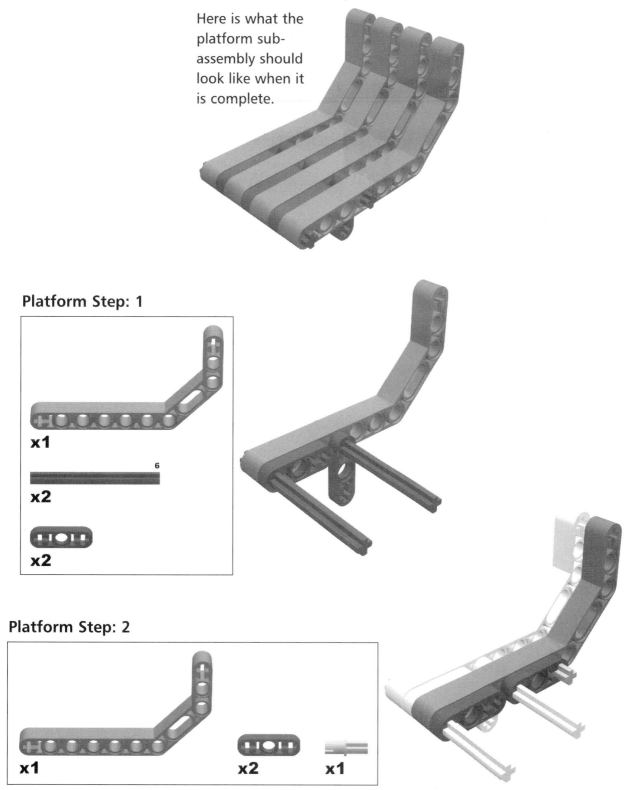

Platform Step: 3

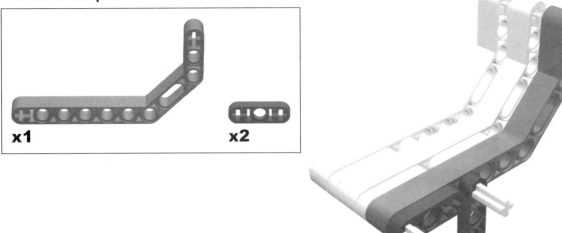

x1   x2

Platform Step: 4

x1

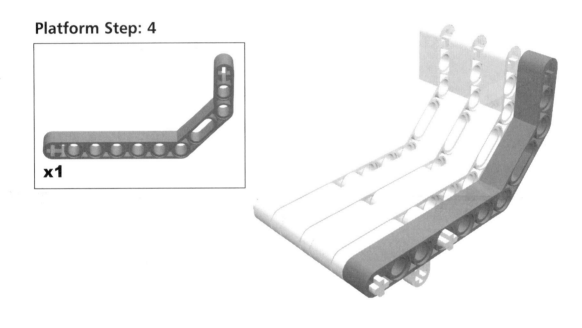

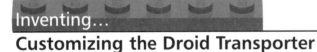

Inventing...

Customizing the Droid Transporter

Try building some small Droids for the Droid Transporter to carry, using the pieces that remain after you have built the robot. LEGO figurines can also be added to the Droid Transporter as passengers, drivers, operators, etc. Whatever you choose to add to the Droid Transporter, make sure the moving ramp is not blocked.

The Ramp

Ramp Step: Complete

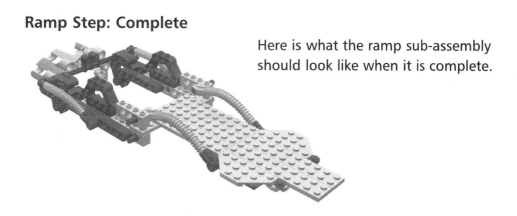

Here is what the ramp sub-assembly should look like when it is complete.

Ramp Step: 1

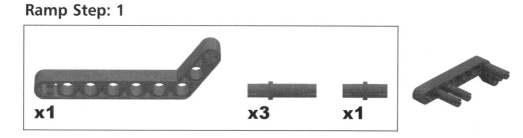

x1   x3   x1

Ramp Step: 2

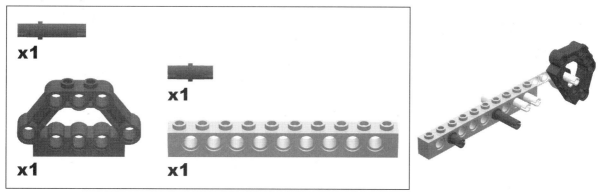

x1

x1

x1   x1

Ramp Step: 3

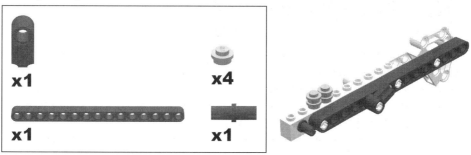

x1   x4

x1   x1

Ramp Step: 4

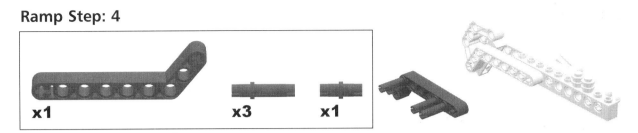

x1   **x3**   **x1**

Ramp Step: 5

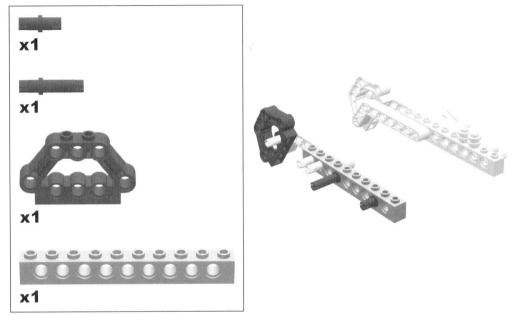

x1

x1

x1

x1

Ramp Step: 6

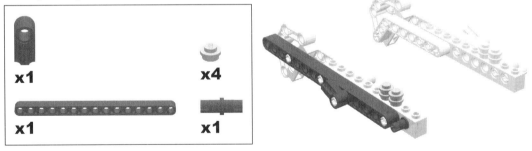

x1   **x4**

x1   **x1**

Ramp Step: 7

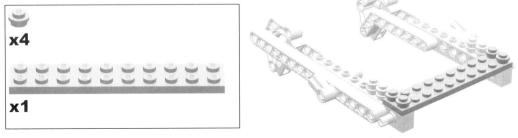

x4

x1

Ramp Step: 8

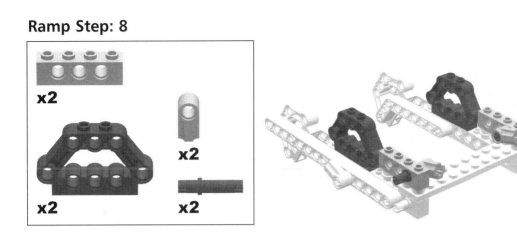

Ramp Step: 9

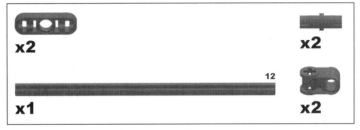

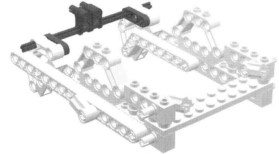

Ramp Step: 10

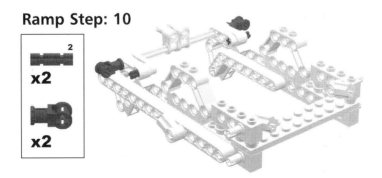

Ramp Step: 11

x3

Ramp Step: 12

x4

Turn the ramp over and attach these pieces as shown.

Ramp Step: 13

x1

x2

x2

x2

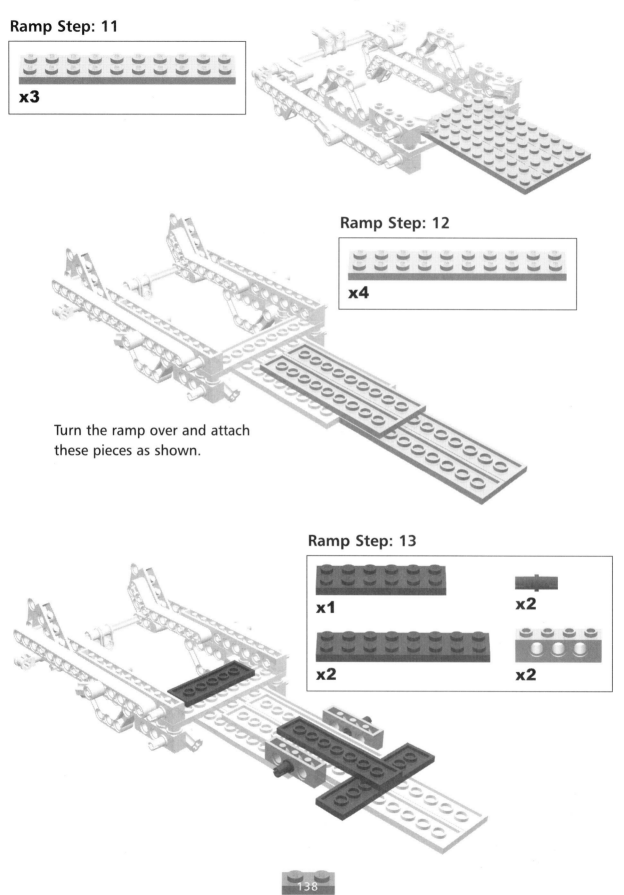

Ramp Step: 14

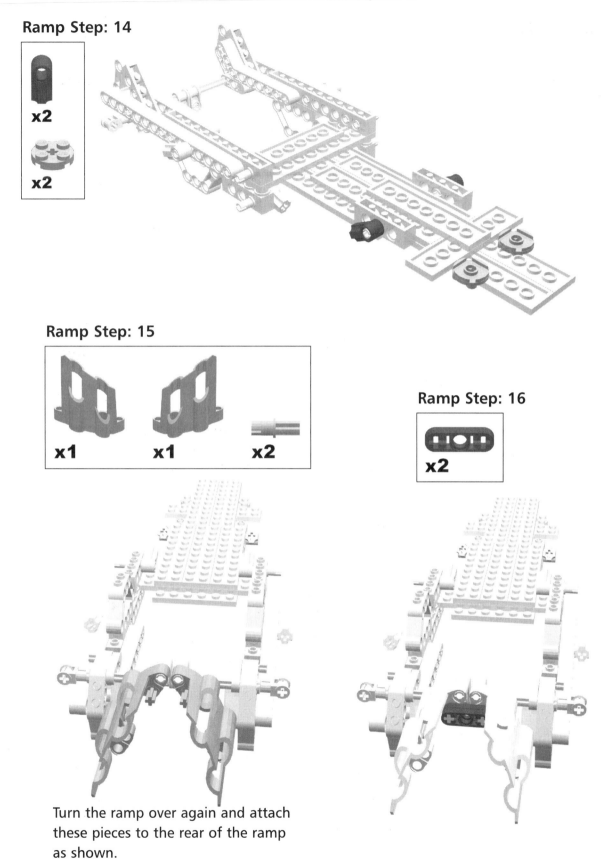

Ramp Step: 15

Ramp Step: 16

Turn the ramp over again and attach
these pieces to the rear of the ramp
as shown.

Ramp Step: 17

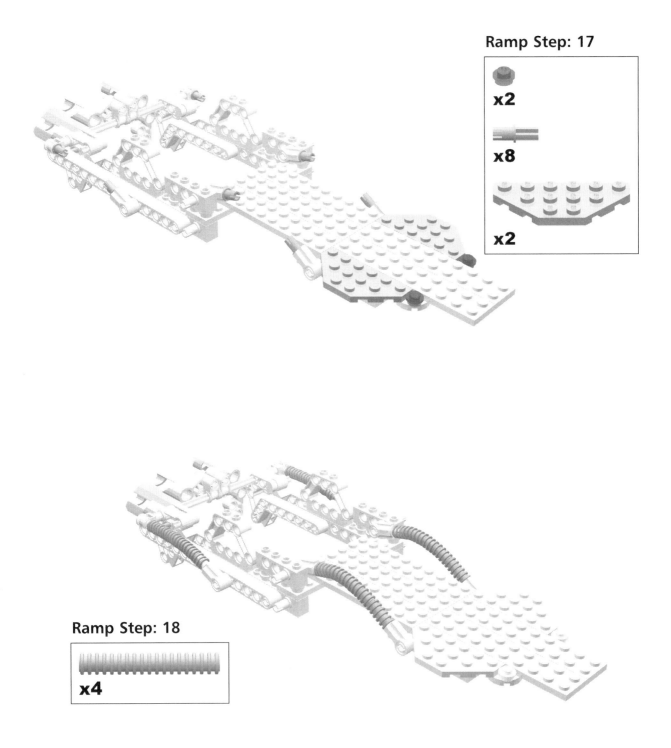

Ramp Step: 18

Final Assembly

You are now ready to take all of the sub-assemblies you have made (the platform and the ramp) and attach them to the body to create the finished Droid Transporter.

Final Assembly Step: Complete

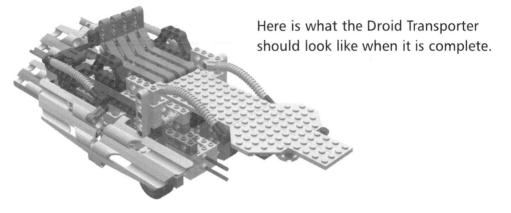

Here is what the Droid Transporter should look like when it is complete.

Final Assembly Step: 1

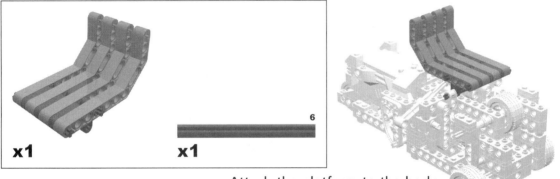

x1   x1   6

Attach the platform to the body with the #6 axle as shown.

Final Assembly Step: 2

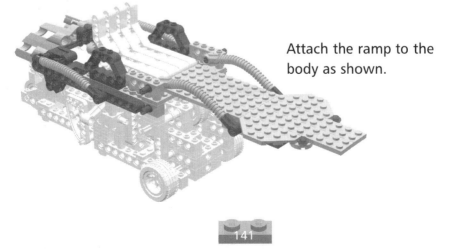

Attach the ramp to the body as shown.

Final Assembly Step: 3

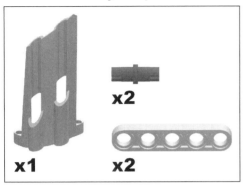

x1   **x2**

x2

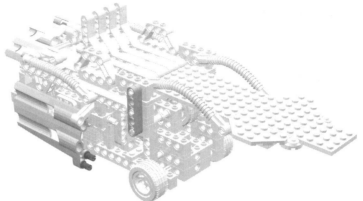

Final Assembly Step: 4

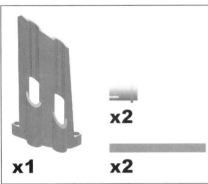

x1   **x2**

x2

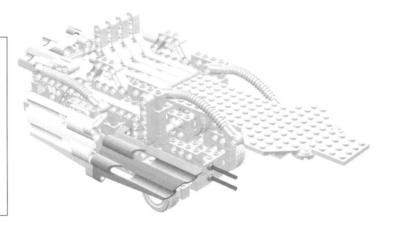

Final Assembly Step: 5

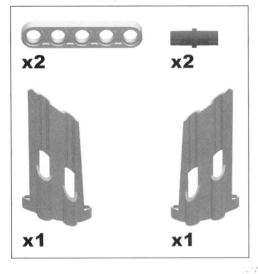

x2   **x2**

x1   **x1**

Turn the Droid
Transporter around
and attach these
pieces to the
opposite side as
shown.

Final Assembly Step: 6

x1   x1

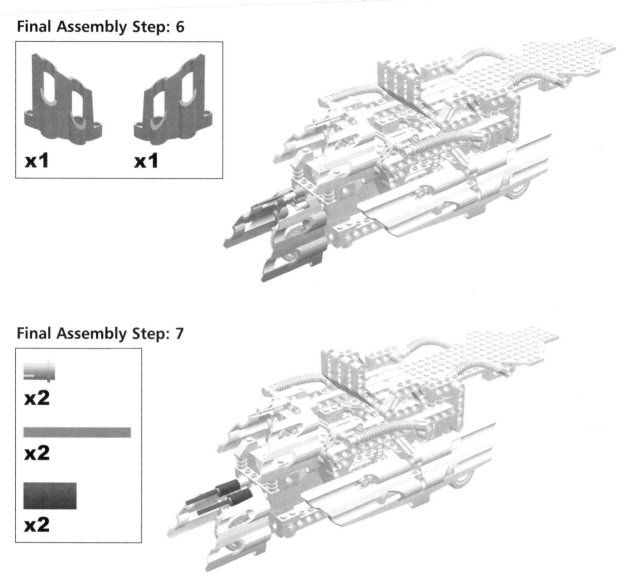

Final Assembly Step: 7

x2

x2

x2

Attach these pieces as shown and your
Droid Transporter is complete!

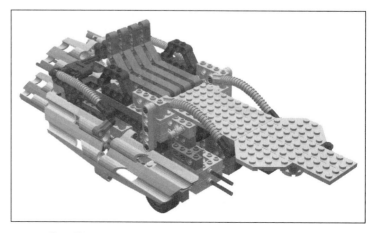

Robot 7

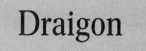

Draigon

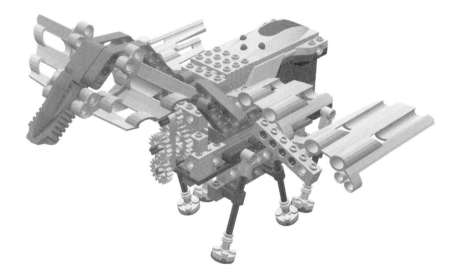

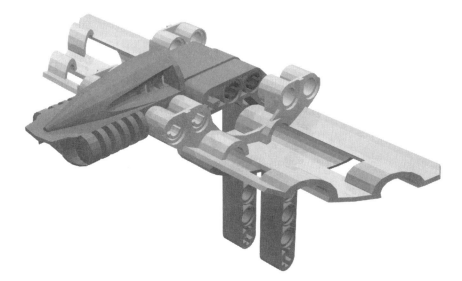

Draigons are fierce flying predators who inhabit several planets throughout the galaxy. They swoop from the sky to attack their prey, carrying off to their lairs to be devoured those creatures unfortunate enough to become their victims. Qui-Gon Jinn and the young Obi-Wan Kenobi encountered them on a nameless watery planet (in *Jedi Apprentice Book 1: The Rising Force*). The transport ship they were traveling on, the *Monument*, crash-landed there after a battle with Togorian pirates. Intent on making the passengers on board the *Monument* their next meal, the Draigons proved to be a challenge even for a Jedi as powerful and experienced as Qui-Gon Jinn, but they were finally defeated.

The LEGO Mindstorms model of the Draigon presented in this chapter is a relatively simple six-legged walker. Once you have built the body for the Draigon, you have a good six-legged base upon which to build your own customized robots.

Micro Scout Programs

You can use any Micro-Scout program to make the Draigon walk.

The Body

The body is the main sub-assembly of the Draigon. Since the six legs are so small on the Draigon (especially when compared to the bipeds presented in this book), we have included them as part of the body itself.

Body Step: Complete

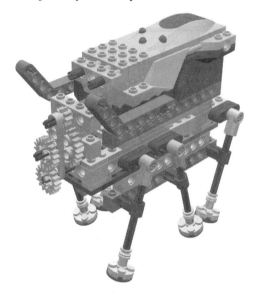

Here is what the body sub-assembly should look like when it is complete.

Body Step: 1

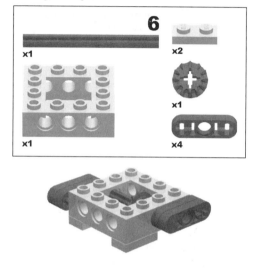

Body Step: 2

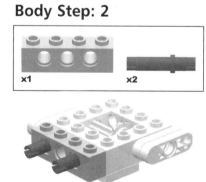

Body Step: 3

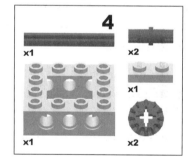

Body Step: 4

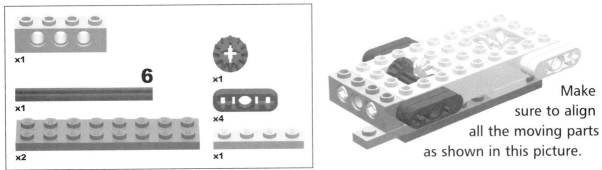

Make sure to align all the moving parts as shown in this picture.

Body Step: 5

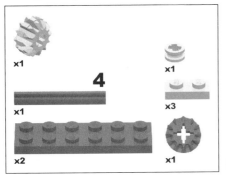

Attach the gears and the small pulley wheel with the #4 axle as shown.

#4 axle

Body Step: 6

Make sure to align all the moving parts as shown in this picture.

Body Step: 7

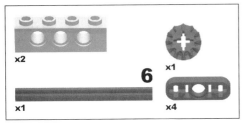

Body Step: 8

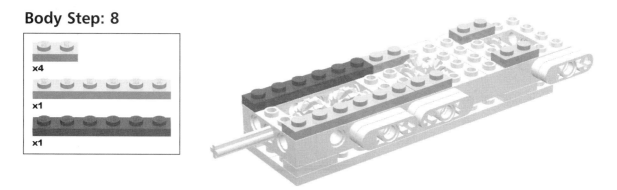

Body Step: 9

Body Step: 10

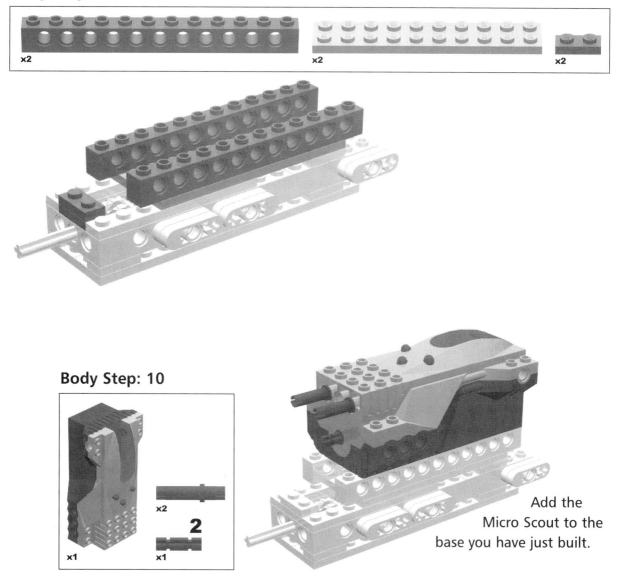

Add the
Micro Scout to the
base you have just built.

Body Step: 11

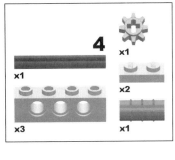

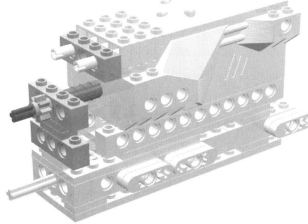

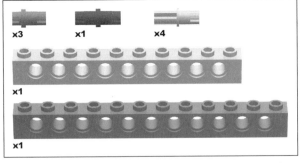

Body Step: 12

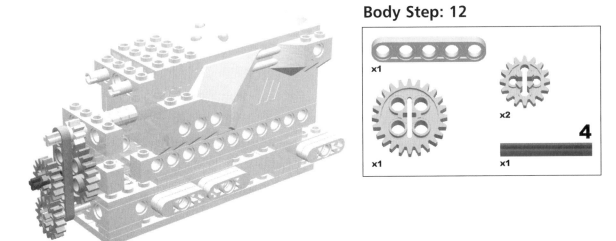

Body Step: 13

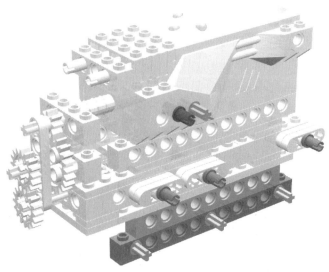

You will now begin construction of the Draigon's left legs.

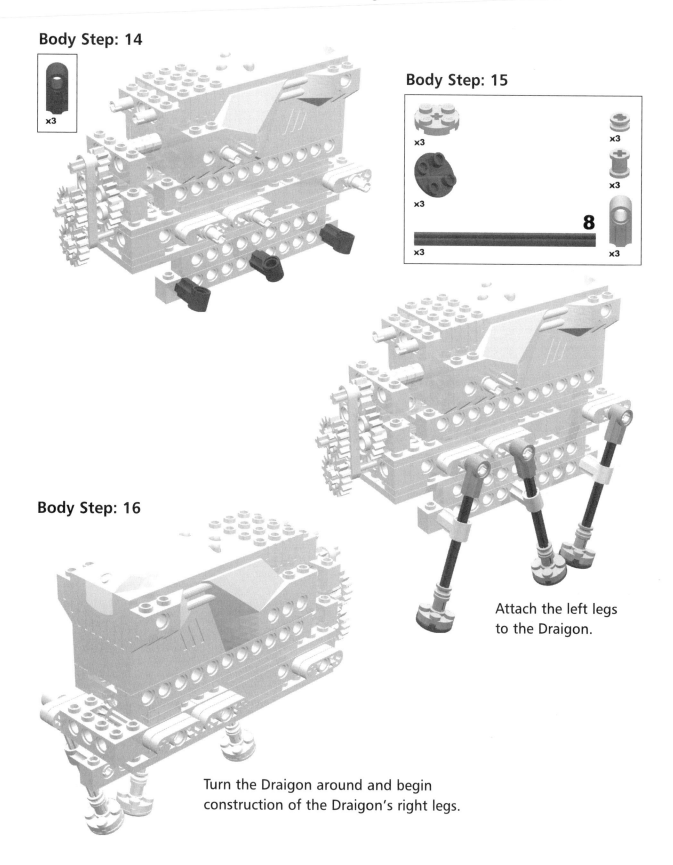

Body Step: 14

x3

Body Step: 15

x3

x3

x3

x3

8

x3

x3

Attach the left legs
to the Draigon.

Body Step: 16

Turn the Draigon around and begin
construction of the Draigon's right legs.

Body Step: 17

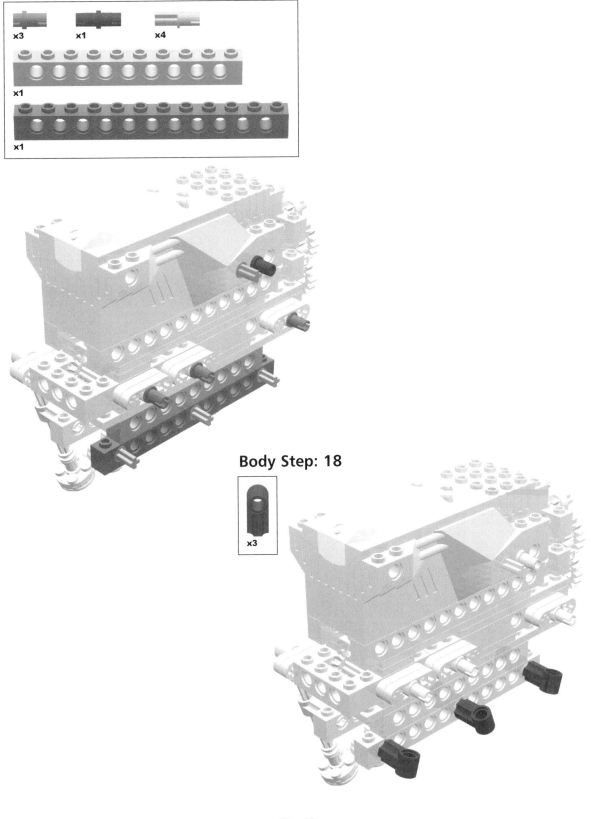

Body Step: 18

Body Step: 19

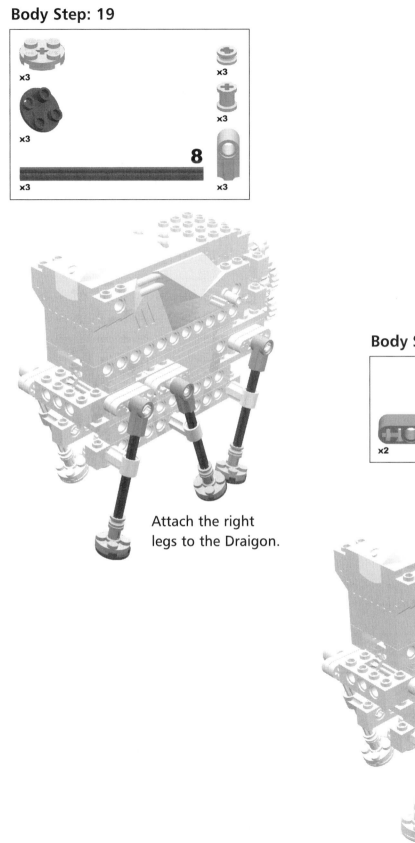

Attach the right
legs to the Draigon.

Body Step: 20

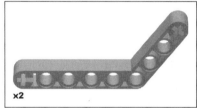

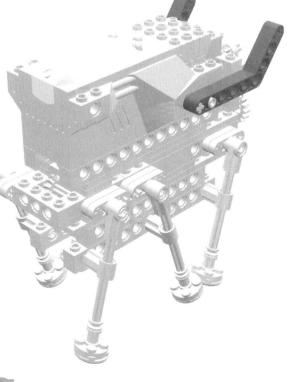

The Right Wing

Right Wing Step: Complete

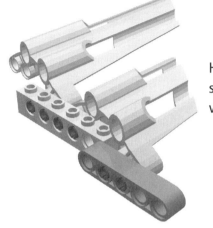

Here is what the right wing sub-assembly should look like when it is complete.

Right Leg Step: 1

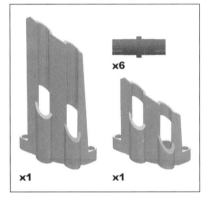

x6

x1   x1

Right Leg Step: 2

x1   x1

The Left Wing

Left Leg Step: Complete

Here is what the left wing sub-assembly should look like when it is complete.

Left Leg Step: 1

x6

x1

x1

Left Leg Step: 2

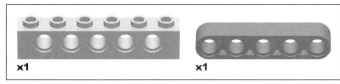

x1

x1

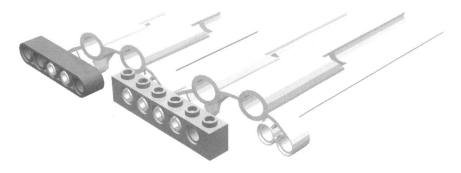

The Head

Head Step: Complete

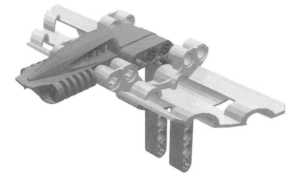

Here is what the head sub-assembly should look like when it is complete.

Inventing...
Customizing the Wings and Head

The wings and head of Draigon are purely decorative. You can make your own wing and head designs, or even make a completely new creature by adding other features to the body sub-assembly, such as a tail, arms, or claws.

Head Step: 1

Head Step: 2

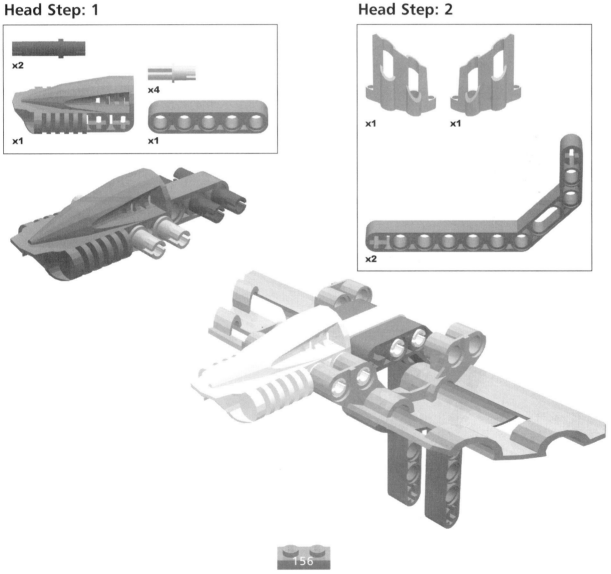

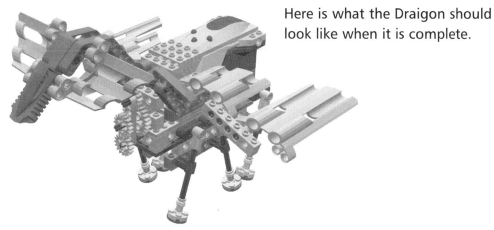

Final Assembly

You are now ready to take all of the sub-assemblies you have made (the right and left wings and the head) and attach them to the body sub-assembly to create the finished Draigon.

Final Assembly Step: Complete

Here is what the Draigon should look like when it is complete.

Final Assembly Step: 1

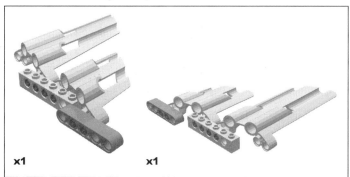

x1   x1

Attach the two wing sub-assemblies to the body as shown.

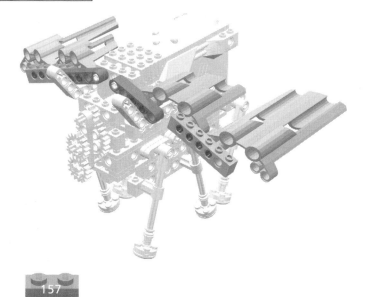

Final Assembly Step: 2

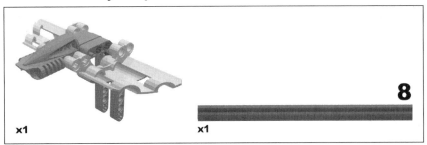

x1   x1   **8**

Attach the head sub-assembly
to the body as shown, and the
Draigon is complete!

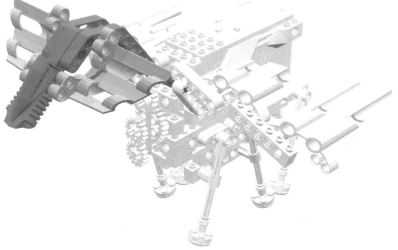

Robot 8

Scorpion Assassin Droid

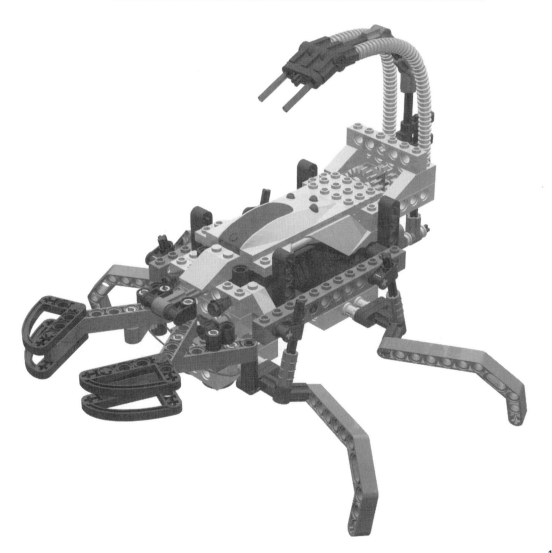

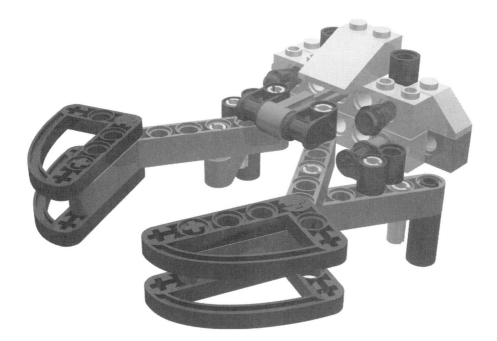

The Scorpion Assassin Droids are legendary, mentioned in hushed whispers in many a cantina and space port throughout the galaxy. No one is certain who controls them–some say the Hutts, others the Empire, and some even believe they are their own masters. Nobody knows for sure, but what is known is that the appearance of a Scorpion Assassin Droid is a sure sign that death is not far behind.

Merely the size of a small dog, a Scorpion Assassin Droid is nevertheless a silent and deadly hunter. Tracking its intended target with a variety of sensors, it moves stealthily to within striking distance. The final killing blow is delivered with a lightning fast flick of its barbed tail, injecting a deadly poison into the victim to which there is no known antidote. Once it acquires a target, it has never been known to fail.

The Scorpion Assassin Droid requires two pieces that are not found in the Dark Side Developer's Kit: two small rubber bands. These should be easy enough for you to find at home or your local corner store, although you may have to do some experimenting to find which size works best (we recommend fairly small rubber bands like the ones orthodontists use for braces if you can find them). You will use these in the final step of the robot.

Bricks & Chips...

Micro Scout Programs

To run the Scorpion Assassin Droid, set the Micro Scout to program #4. Cover the light sensor with your hand and press **RUN**. When the Micro Scout stops beeping, remove your hand and the Scorpion Assassin Droid will begin moving!

The Body

The body of the Scorpion Assassin Droid includes the four legs that it uses to walk in addition to the more traditional Micro Scout body that you might be familiar with by now.

Body Step: Complete

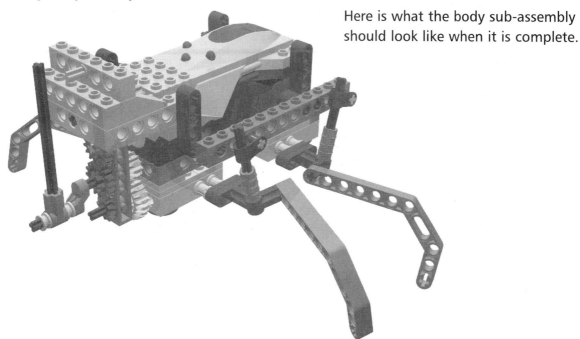

Here is what the body sub-assembly should look like when it is complete.

Body Step: 1

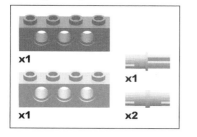

Body Step: 2

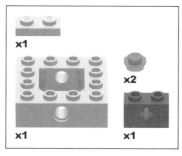

Body Step: 3

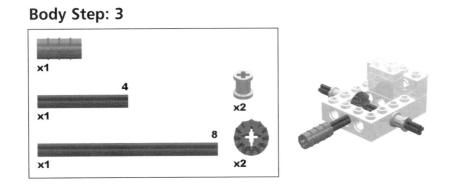

Body Step: 4

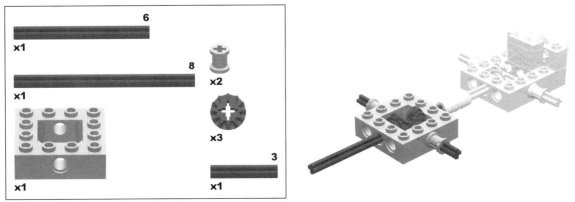

Body Step: 5

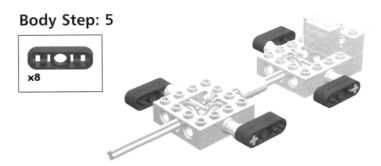

Carefully look at this picture before attaching the 1x3 liftarms in this step. It is very important that the four sets of 1x3 liftarms are pointing in the direction shown here and that they are attached exactly as shown.

Body Step: 6

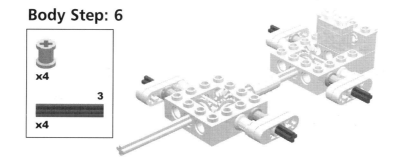

Body Step: 7

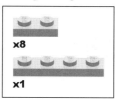

Turn the body over and add these pieces as shown.

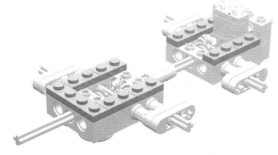

Body Step: 8

Turn the body back over and add these pieces as shown.

Body Step: 9

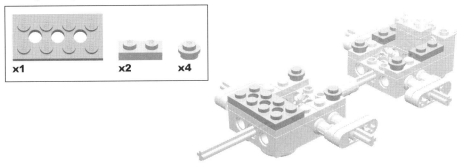

Body Step: 10

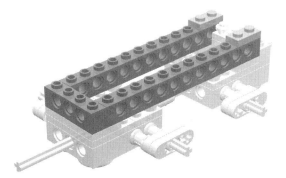

Body Step: 11

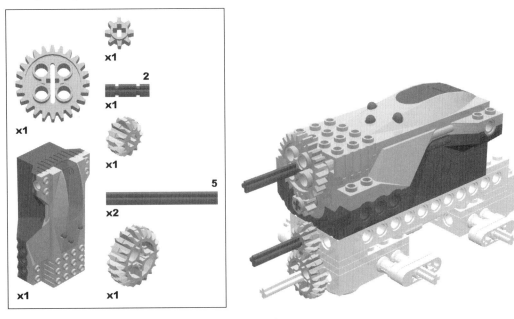

Body Step: 12

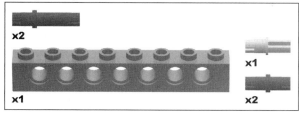

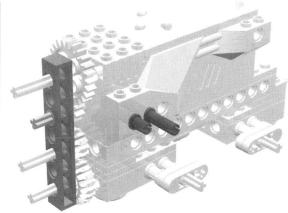

Body Step: 13

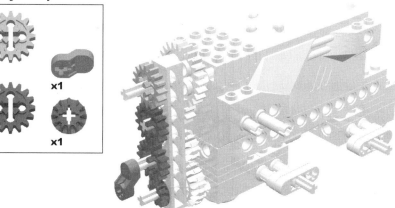

Body Step: 14

Add these pieces to both sides of the body as shown.

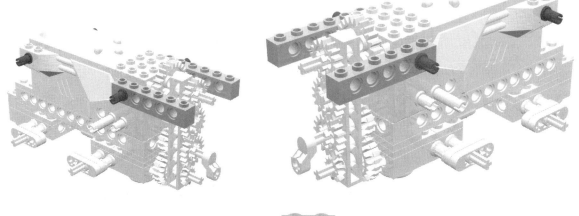

Body Step: 15

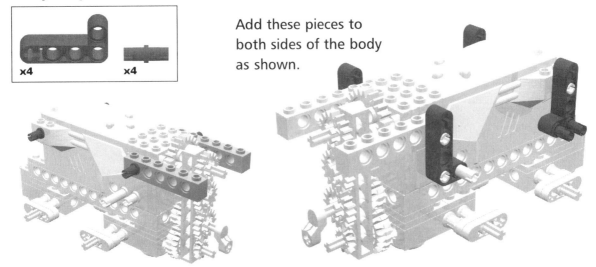

x4 x4

Add these pieces to both sides of the body as shown.

Body Step: 16

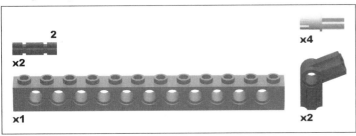

2

x2

x4

x1

x2

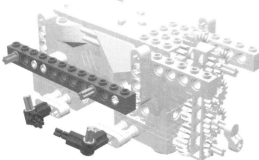

Add these pieces to both sides of the body as shown.

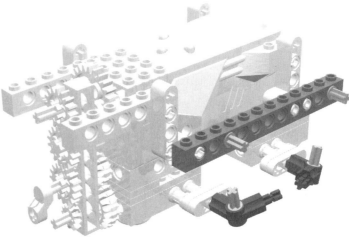

Body Step: 17

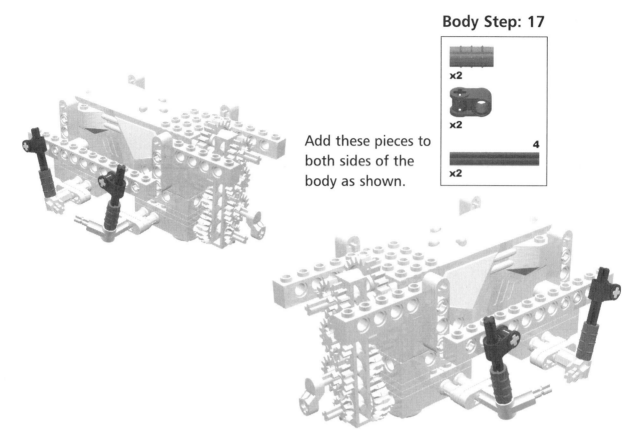

Add these pieces to both sides of the body as shown.

x2

x2

4

x2

Body Step: 18

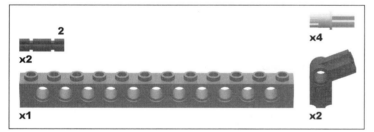

2

x2

x1

x4

x2

Add these pieces to both sides of the body as shown.

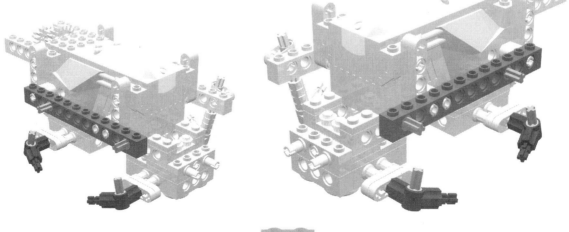

Body Step: 19

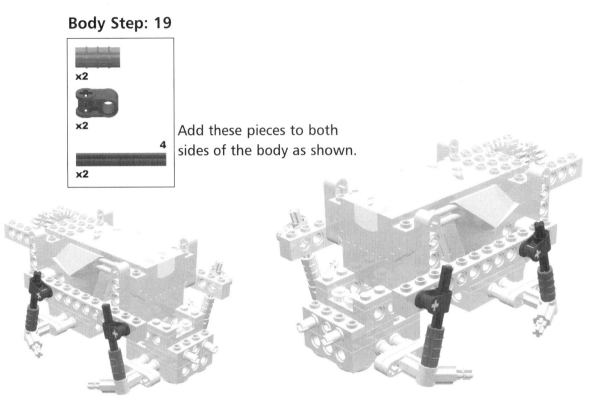

Add these pieces to both sides of the body as shown.

Inventing...

Customizing the Scorpion Assassin Droid

Because the Scorpion Battle Droid has four legs, it is very stable compared to the bipeds presented in this book. This means that there are more options available to you for customizing it. For example, try to equip the Scorpion Assassin Droid with wings on its sides. If you try this, make sure the wings do not interfere with the moving parts already present (the legs, the tail, and the claws). You could also try different variations on the tail, but make sure not to make it too heavy or it will not move properly.

Body Step: 20

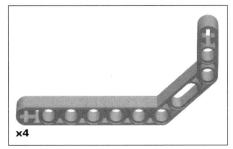

x4

Add the legs to both sides
of the body as shown.

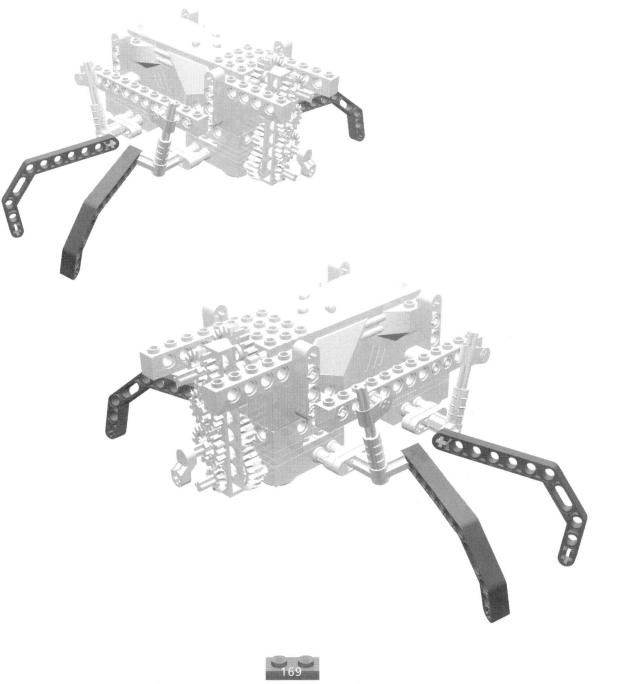

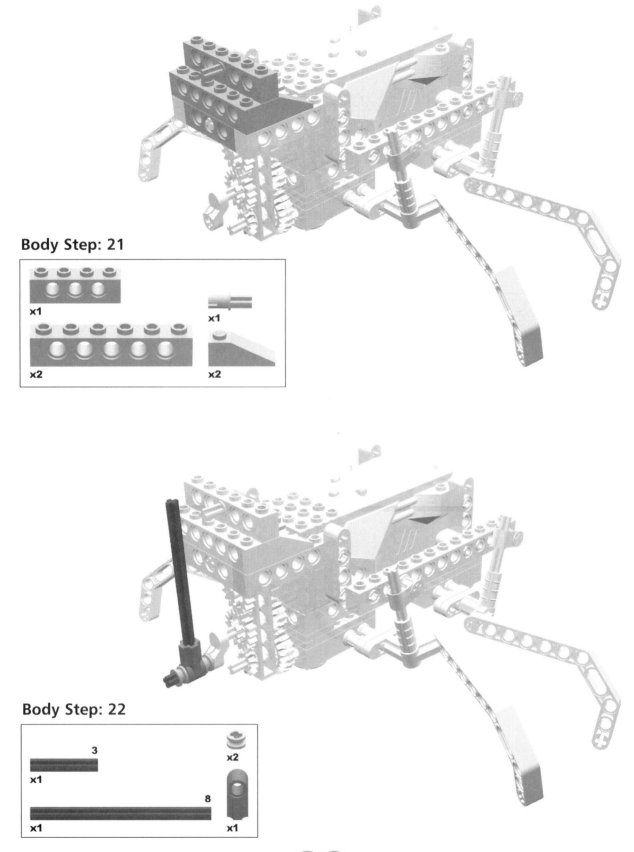

Body Step: 21

x1
x2
x1
x2

Body Step: 22

3
x1
8
x1
x2
x1

The Tail

Tail Step: Complete

Here is what the tail should
look like when it is complete.

Tail Step: 1

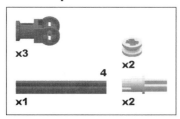

Pass the #4 axle through
these parts as shown.

Tail Step: 2

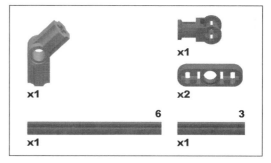

Use the #3 axle
for the top piece
of this step.

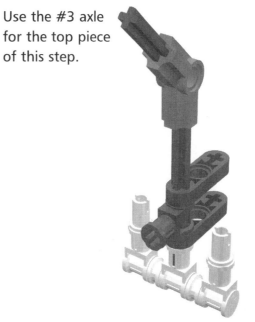

Tail Step: 3

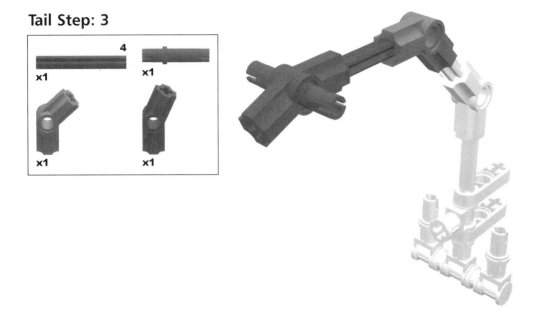

Tail Step: 4

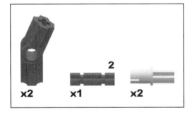

Tail Step: 5

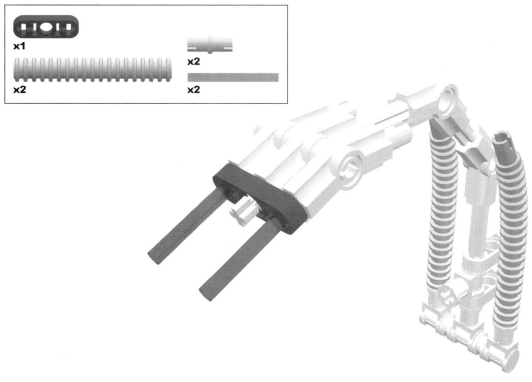

Tail Step: 6

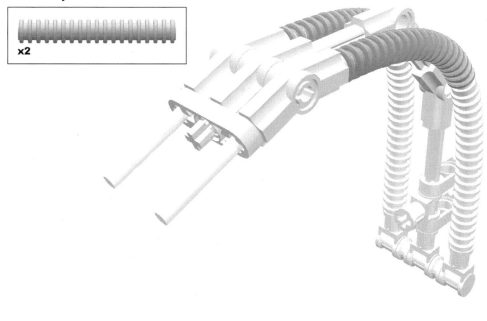

The Claws

Claws Step: Complete

Here is what the claws should look like when they are complete.

Claws Step: 1

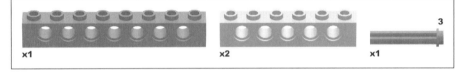

x1   x2   3   x1

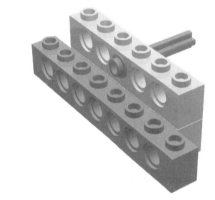

Claws Step: 2

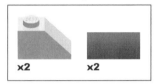

x2   x2

Turn the claws around and add these pieces as shown.

Claws Step: 3

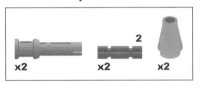

x2   2   x2   x2

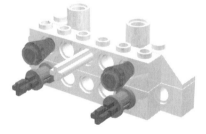

Claws Step: 4

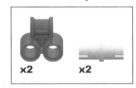

x2   x2

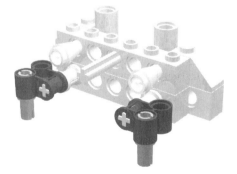

Claws Step: 5

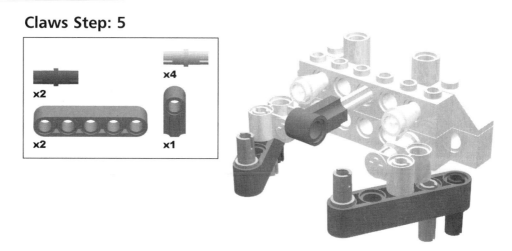

Claws Step: 6

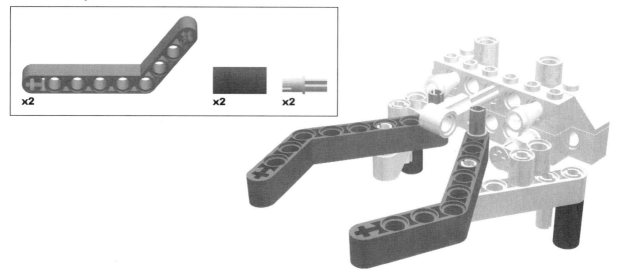

Claws Step: 7

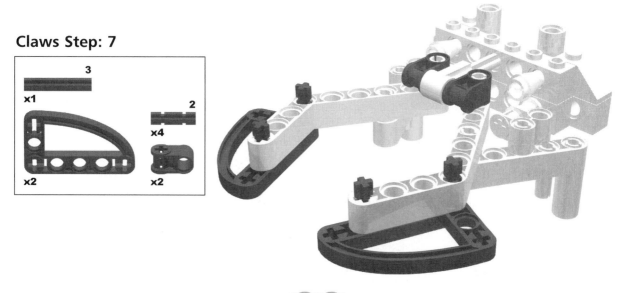

Claws Step: 8

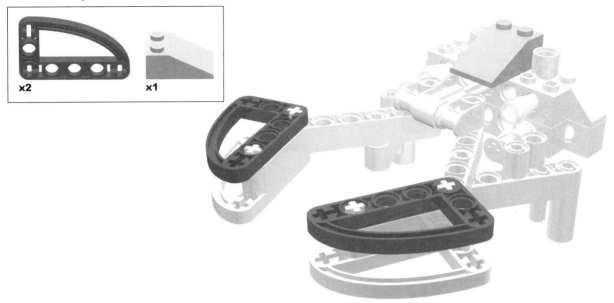

Final Assembly

You are now ready to take all of the sub-assemblies you have made (the tail and claws sub-assemblies) and attach them to the body sub-assembly to create the finished Scorpion Assassin Droid.

Final Assembly Step: Complete

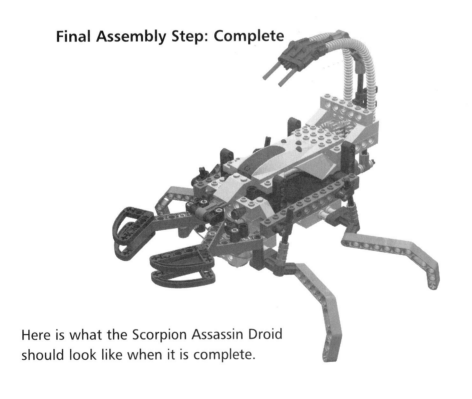

Here is what the Scorpion Assassin Droid should look like when it is complete.

Final Assembly Step: 1

Attach the tail to the body as shown.

x1

x1

Final Assembly Step: 2

x1

Attach the claws to the body as shown.

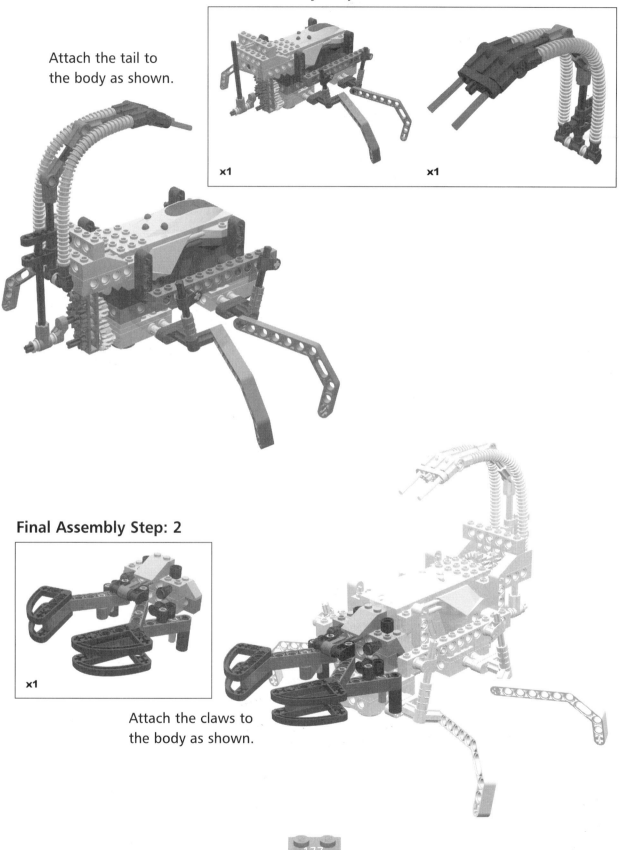

Final Assembly Step: 3

x2

Attach two rubber bands to the claws
as shown. Once this is done, the
Scorpion Assassin Droid is complete!

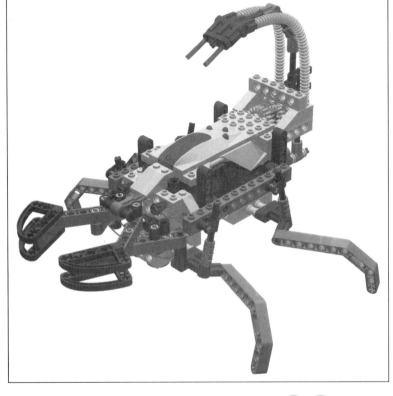

Robot 9

Ludic Ordinance Unit (LOU)

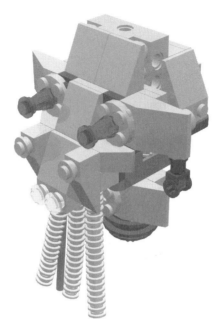

Found in most Imperial barracks across the galaxies, Ludic Ordinance Units, called LOUs, are usually put together by Imperial engineers from bits and pieces, to create entertainment for the troops. Most end up serving as moving targets for Imperial Stormtrooper blaster practice after a particularly poor performance. Some, however, turn out to be hilarious comedians and gifted musicians, and are presented from time to time to the Imperial officers for their amusement. It is rumored that Darth Vader has a personal LOU quartet that plays private shows for him, but many people dismiss the thought of the Emperor's right-hand man enjoying music as preposterous.

Bricks & Chips...

Micro Scout Programs

A LOU can use any of the Micro Scout programs. However, you will find that adjusting the arm that strums the guitar is a delicate task, so you might want to use program #1 for a constant movement.

The Body

Begin by building the body sub-assembly. You will then attach all of the other sub-assemblies to the body.

Body Step: Complete

Here is what the body sub-assembly should look like when it is complete.

Body Step: 1

Body Step: 2

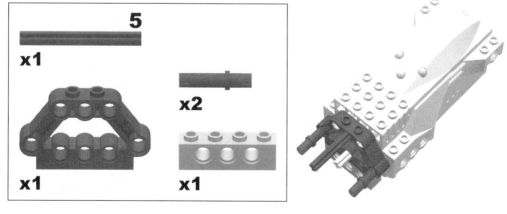

Body Step: 3

x1 **x1**

Body Step: 4

x1 **x1** **x1**

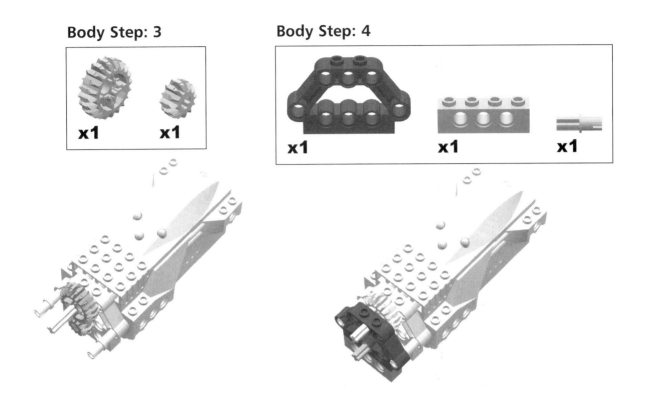

Body Step: 5

x1 **x1**

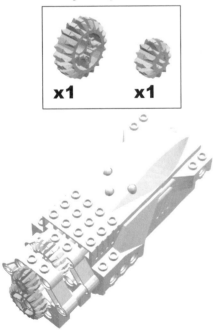

Body Step: 6

x4 Turn the body over and add these pieces as shown.

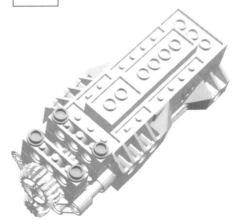

Body Step: 7

x1   x4

Body Step: 8

x4

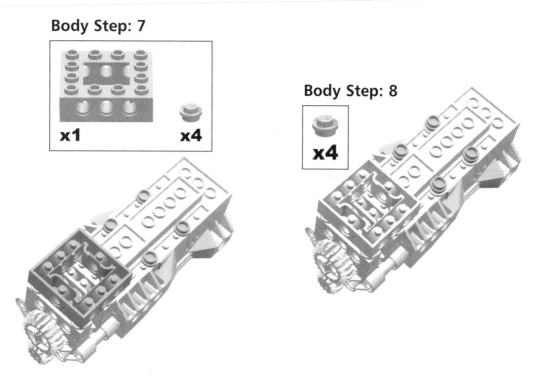

Body Step: 9

x1

x1

12

x1

x1

Body Step: 10

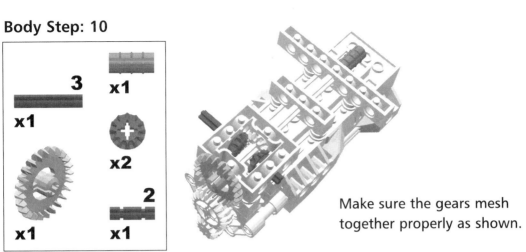

Make sure the gears mesh together properly as shown.

Body Step: 11

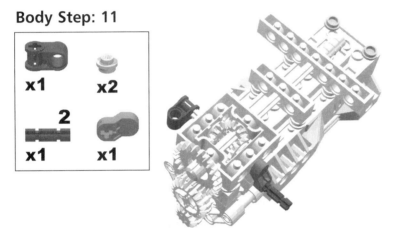

Body Step: 12

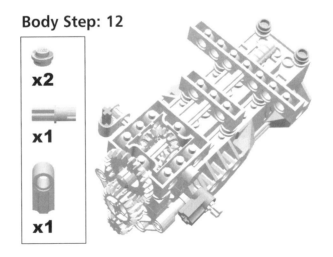

Body Step: 13

x1

x1

x1

5
x1

2
x1

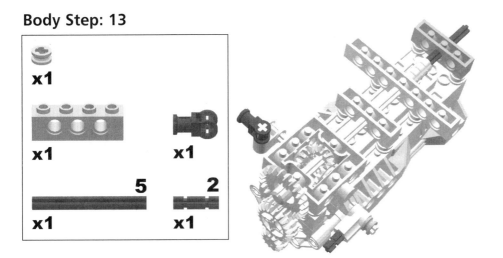

Body Step: 14

x1

x1

x1

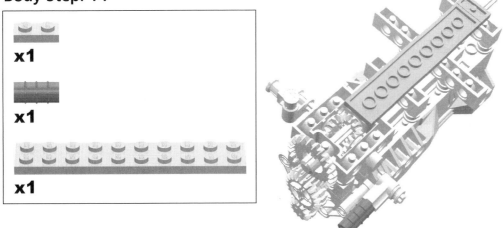

Body Step: 15

x2

x1

x1

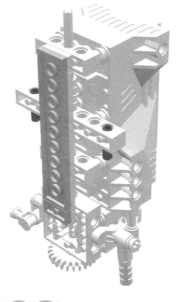

Body Step: 16

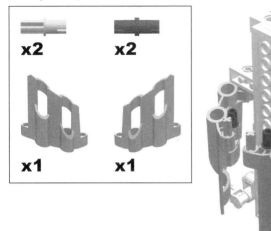

x2 **x2**

x1 **x1**

Body Step: 17

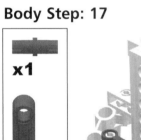

x1

x1

Body Step: 18

2

x1

x1

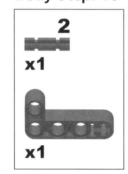

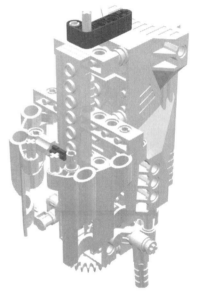

Body Step: 19

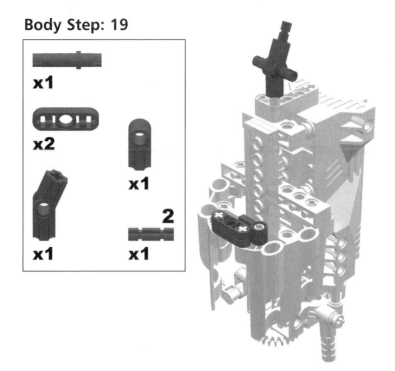

Body Step: 20

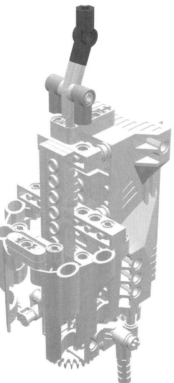

The Left Leg

Left Leg Step: Complete

Here is what the left leg sub-assembly should look like when it is complete. When the LOU is completed, the Micro Scout will enable the LOU to tap its left foot on the floor.

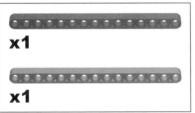

Left Leg Step: 1

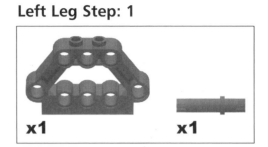

x1   x1

Left Leg Step: 2

x3

Left Leg Step: 3

x1

x1

Left Leg Step: 4

x1

x3

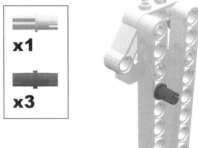

Left Leg Step: 5

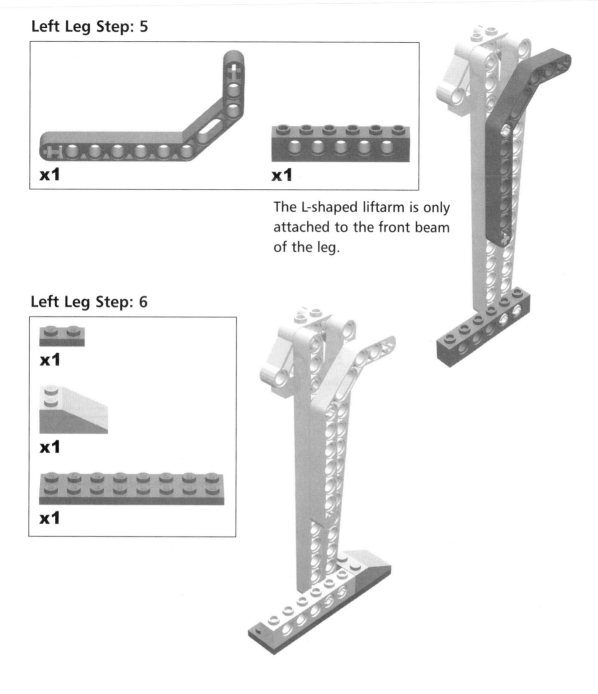

x1   **x1**

The L-shaped liftarm is only attached to the front beam of the leg.

Left Leg Step: 6

x1

x1

x1

Left Leg Step: 7

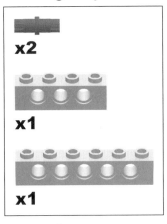

x2

x1

x1

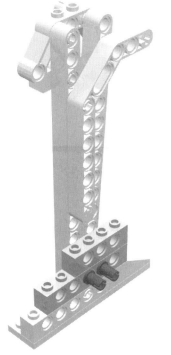

Left Leg Step: 8

x1

3
x1

Left Leg Step: 9

x1

x2

2
x1

Left Leg Step: 10

x1

Left Leg Step: 11

x1

Left Leg Step: 12

x1

x2

x1

Left Leg Step: 13

x1

x1

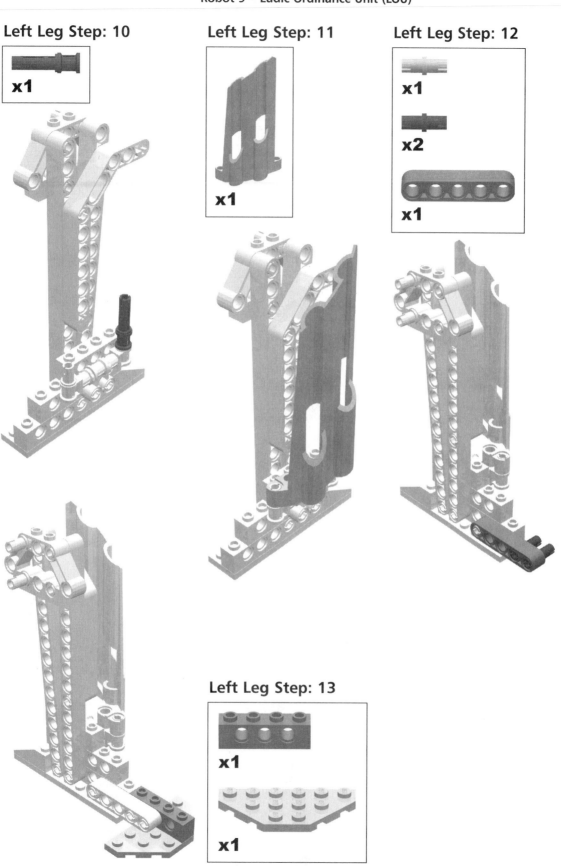

Left Leg Step: 14

x3

x1

Left Leg Step: 15

x1

x2

Left Leg Step: 16

x1

Turn the left leg around and begin work on the rear of the leg.

Left Leg Step: 17

x2

x1

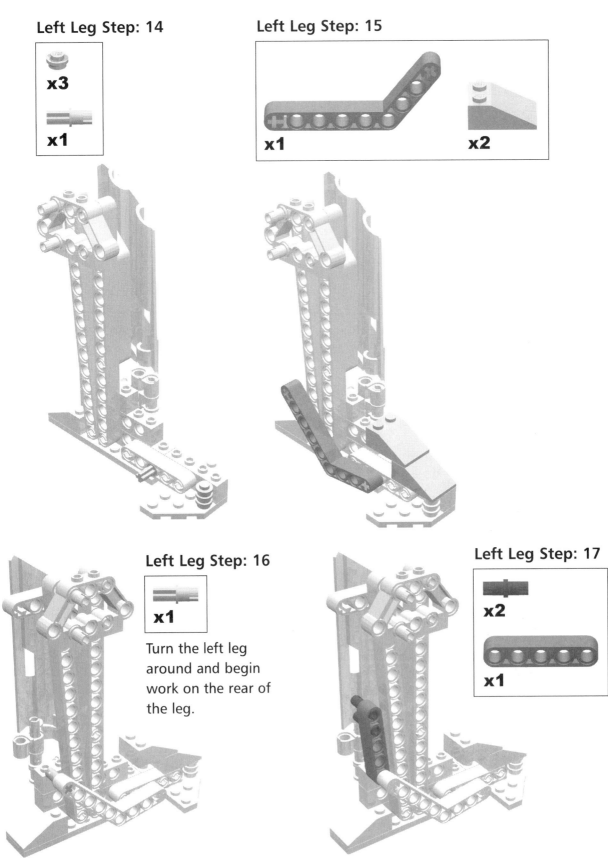

Left Leg Step: 18

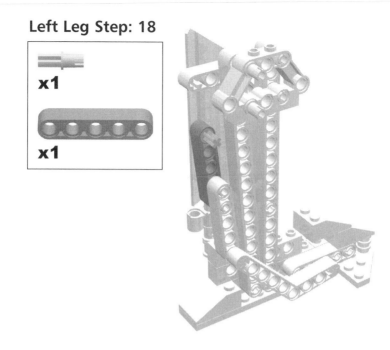

x1

x1

The Right Leg

The right leg looks identical to the left leg, but its construction and purpose are somewhat different. The right leg lifts the entire LOU robot up and down.

Right Leg Step: Complete

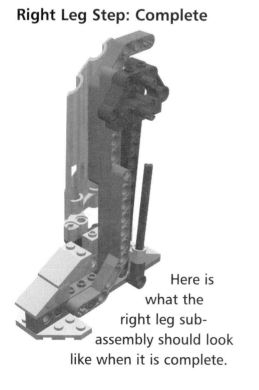

Here is what the right leg sub-assembly should look like when it is complete.

Right Leg Step: 1

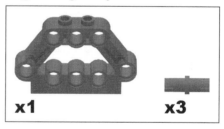

x1   x3

Right Leg Step: 2

x1

x1

Right Leg Step: 3

x2

x1

x2

Right Leg Step: 4

x1

x1

x1

x1

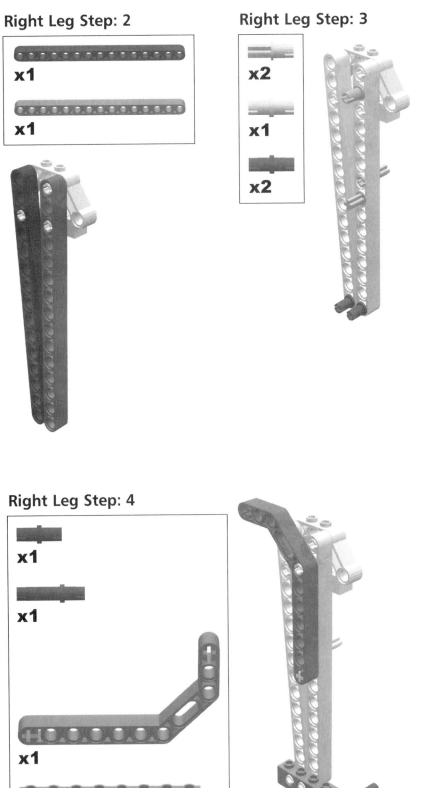

Right Leg Step: 5

x1

x1

Right Leg Step: 6

x1

x2

Right Leg Step: 7

x2

x1

x1

x1

x1

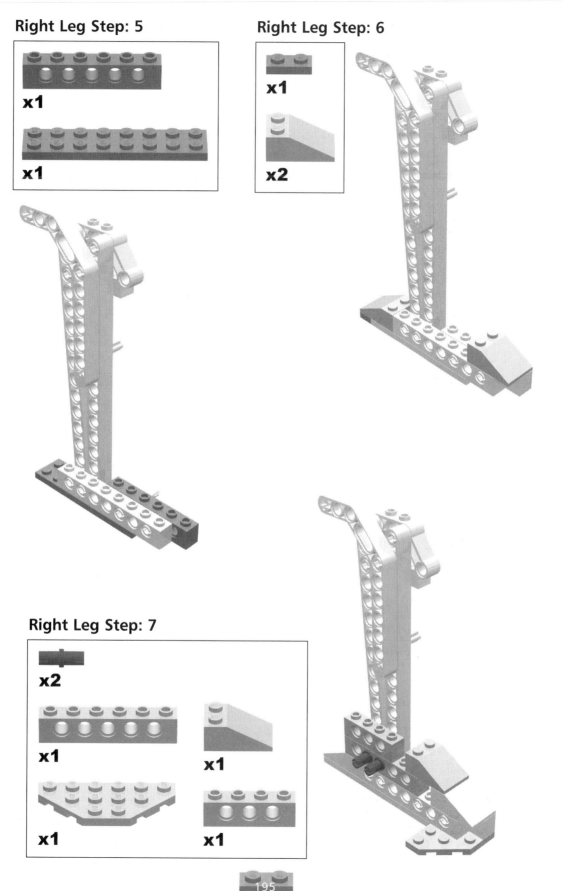

Right Leg Step: 8

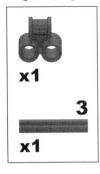

x1

3

x1

Right Leg Step: 9

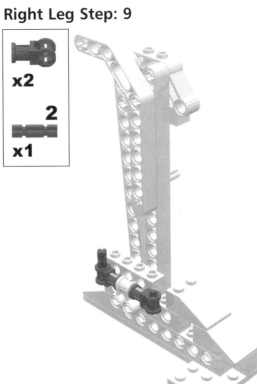

x2

2

x1

Right Leg Step: 10

x1

x1

Right Leg Step: 11

x1

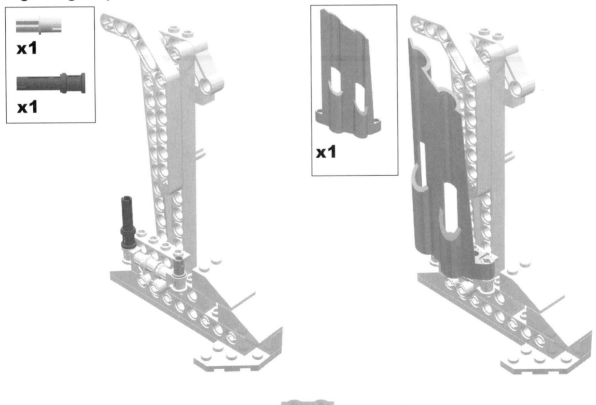

Right Leg Step: 12

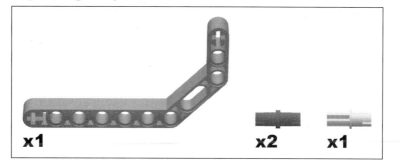

x1   x2   x1

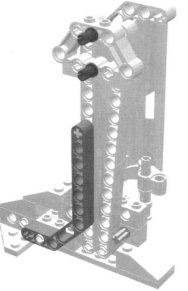

Turn the right leg around
and begin work on the
rear of the leg.

Right Leg Step: 13

x1

x1

8

x1

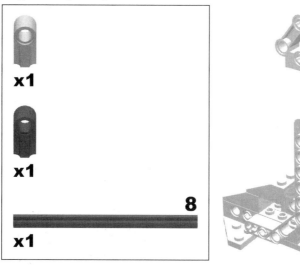

The Head

Head Step: Complete

Here is what the head sub-assembly should look like when it is complete.

Head Step: 1

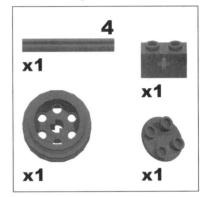

4
x1   x1
x1   x1

Head Step: 2

x2   x2
x1   x1

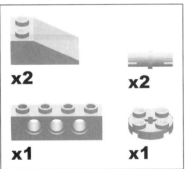

Head Step: 3

x2   x2
x2   x2

Head Step: 4

x2

x1

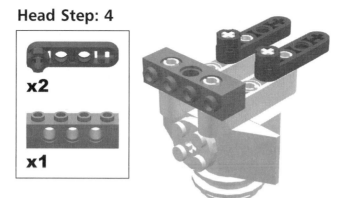

Head Step: 5

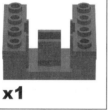

x1

Begin creating the face for the LOU.

Head Step: 6

x2

x1

x2

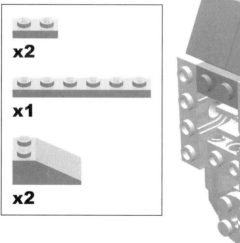

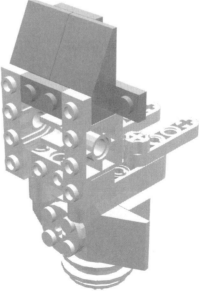

Head Step: 7

Head Step: 8

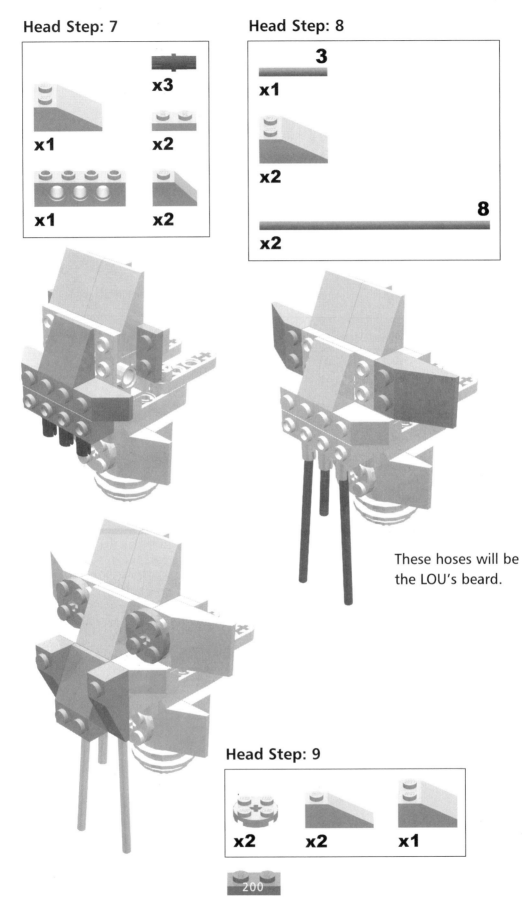

These hoses will be the LOU's beard.

Head Step: 9

Head Step: 10

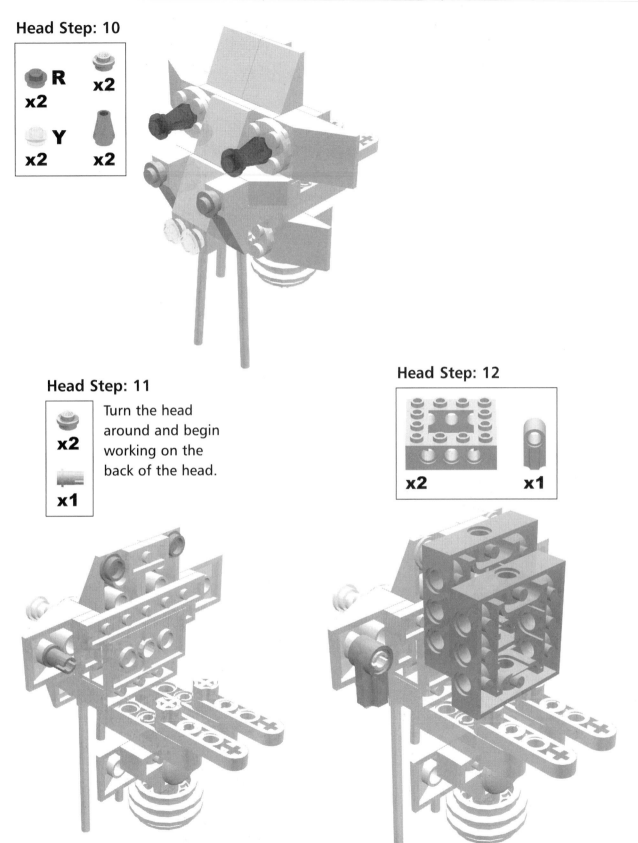

Head Step: 11

Turn the head around and begin working on the back of the head.

Head Step: 12

Head Step: 13

x4

2

x1

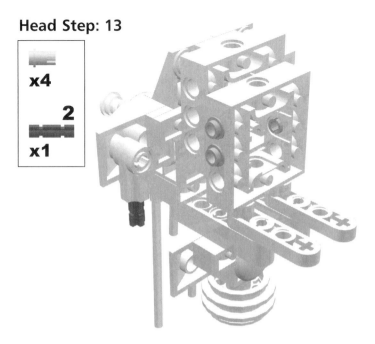

Head Step: 14

x1

x1

x1

x1

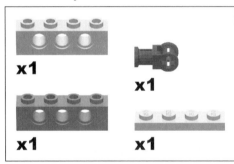

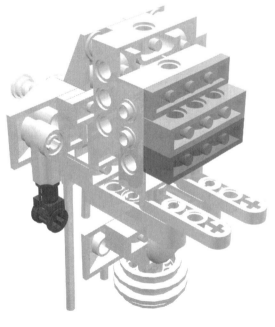

Head Step: 15

x2

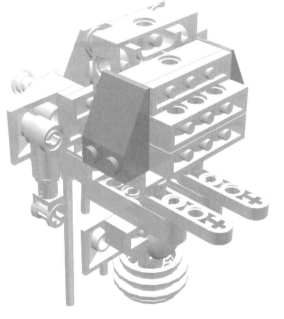

Head Step: 16

x3

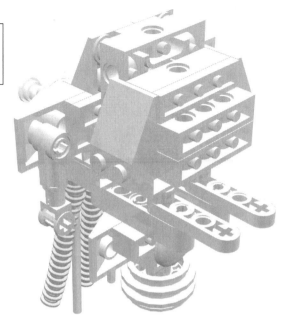

The Left Arm

Left Arm Step: Complete

Here is what the left arm sub-assembly should look like when it is complete.

Left Arm Step: 1

2

x1

x1

x1

x1

Left Arm Step: 2

x2 x1 **3** x1

x1

Left Arm Step: 3

x1 x1

x1 x1

Left Arm Step: 4

x2

R

x2

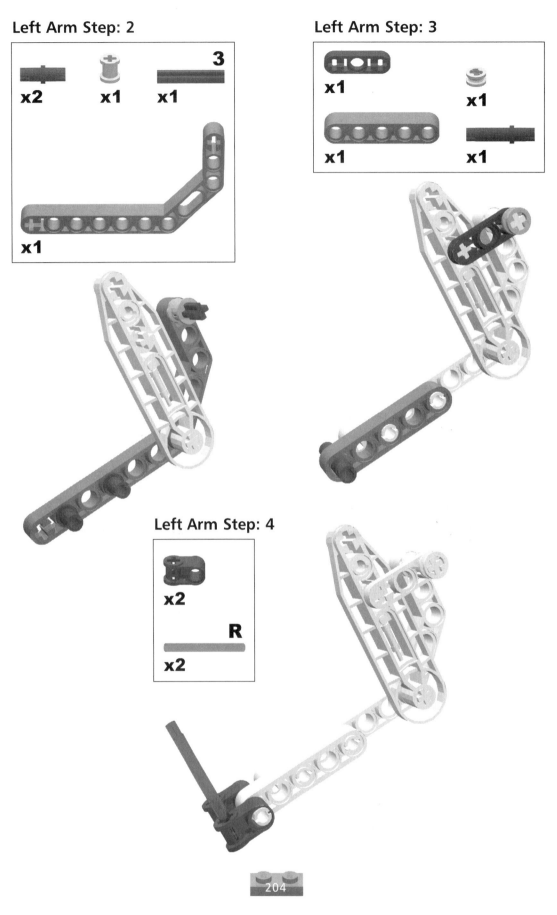

Left Arm Step: 5

x1

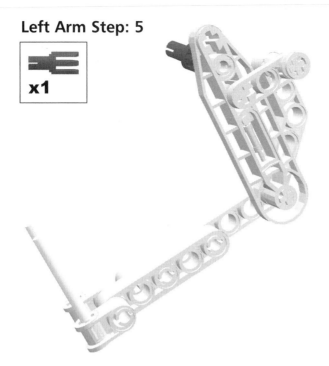

The Right Arm

Right Arm Step: Complete

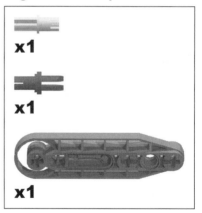

Here is what the right arm sub-assembly should look like when it is complete.

Right Arm Step: 1

x1

x1

x1

Right Arm Step: 2

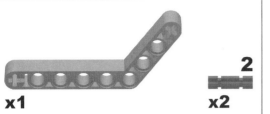

x1   2 x2

Right Arm Step: 3

x1   x1

8 x1   x1

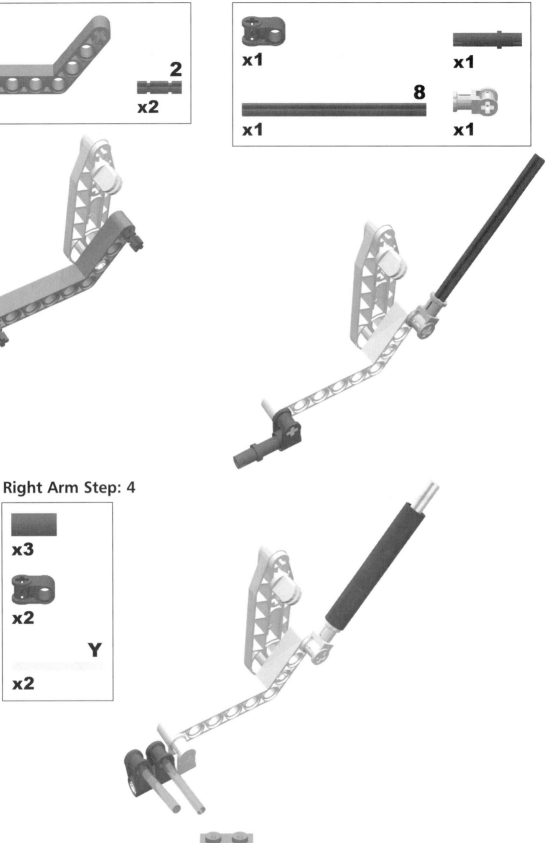

Right Arm Step: 4

x3

x2

Y

x2

Right Arm Step: 5

x5

x1

Attach the five round plates with the studs facing down the length of the arm, as shown.

Right Arm Step: 6

x1

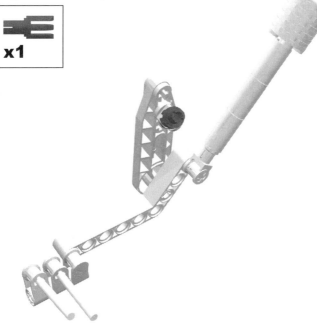

The Guitar

Guitar Step: Complete

Here is what the guitar sub-assembly should look like when it is complete.

Guitar Step: 1

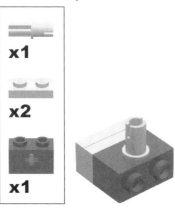

x1

x2

x1

These bricks will attach the back of the guitar to the LOU's body.

Guitar Step: 2

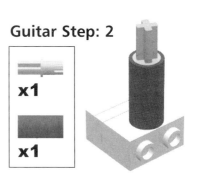

x1

x1

Guitar Step: 3

x1   2

x1   x1

Guitar Step: 4

x2

x1

Guitar Step: 5

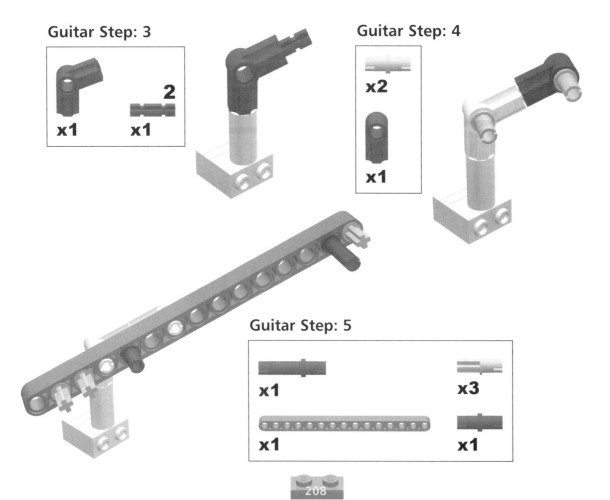

x1   x3

x1   x1

Guitar Step: 6

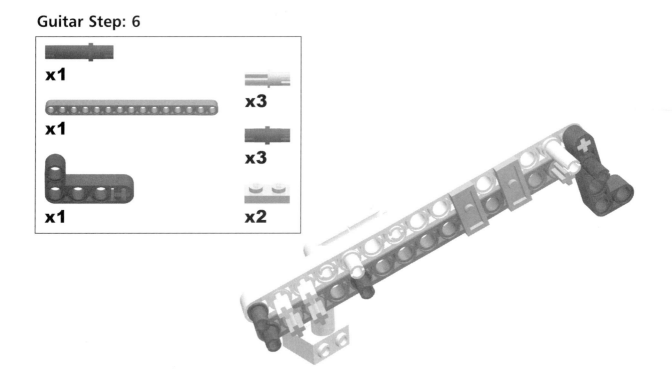

Guitar Step: 7

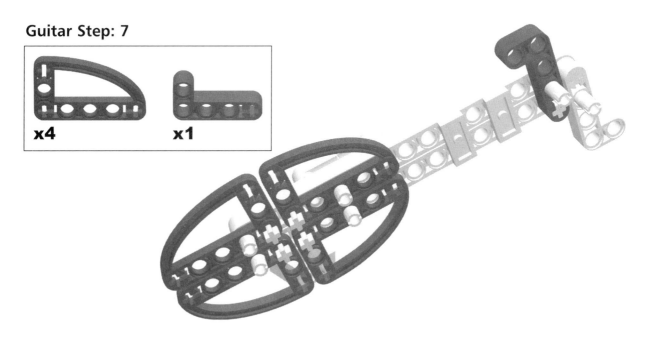

Guitar Step: 8

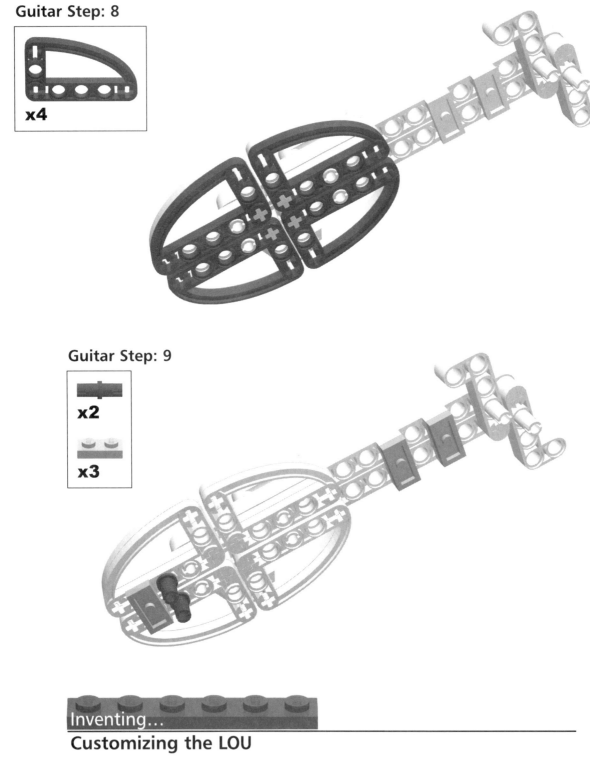

x4

Guitar Step: 9

x2

x3

Inventing...

Customizing the LOU

One easy way to customize the LOU is to add guitar strings to the guitar. You can do this by placing rubber bands or string on the pegs on the front of the guitar. Try building other musical instruments for the LOU to play as well.

Final Assembly

You are now ready to take all of the sub-assemblies you have built (the right and left legs, the right and left arms, the head, and the guitar) and attach them to the body sub-assembly to create the finished LOU robot.

Final Assembly Step: Complete

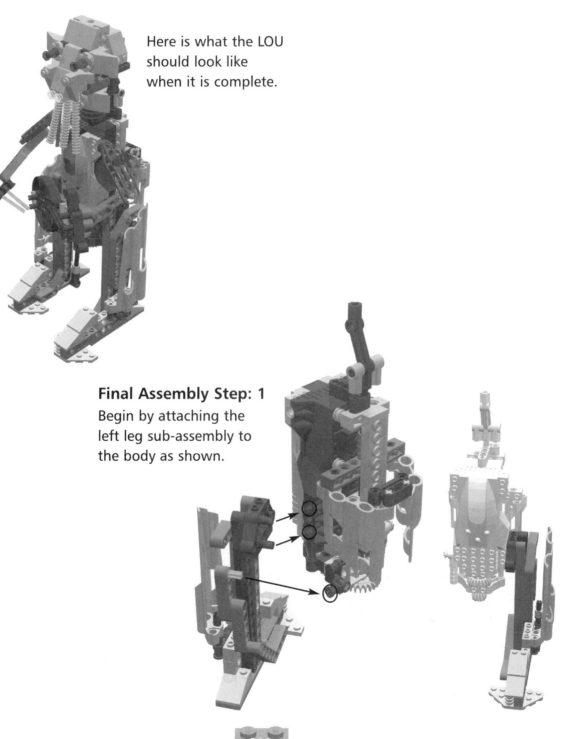

Here is what the LOU should look like when it is complete.

Final Assembly Step: 1

Begin by attaching the left leg sub-assembly to the body as shown.

Final Assembly Step: 2

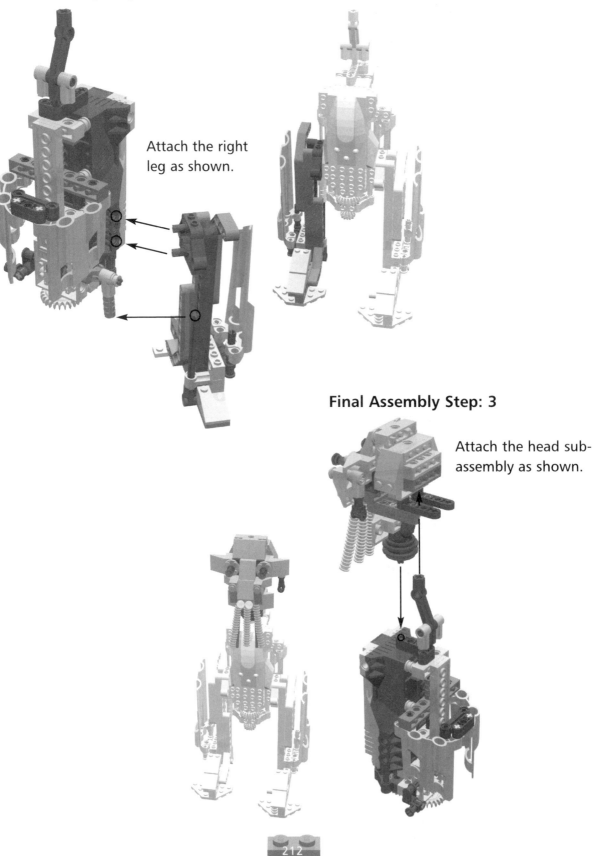

Attach the right leg as shown.

Final Assembly Step: 3

Attach the head sub-assembly as shown.

Final Assembly Step: 4

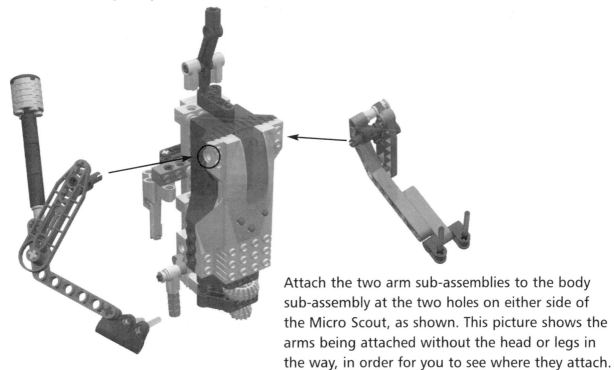

Attach the two arm sub-assemblies to the body sub-assembly at the two holes on either side of the Micro Scout, as shown. This picture shows the arms being attached without the head or legs in the way, in order for you to see where they attach.

Final Assembly Step: 5

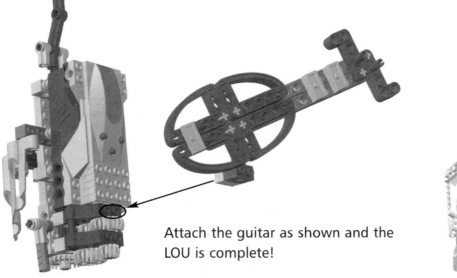

Attach the guitar as shown and the LOU is complete!

Robot 10

Imperial Hound

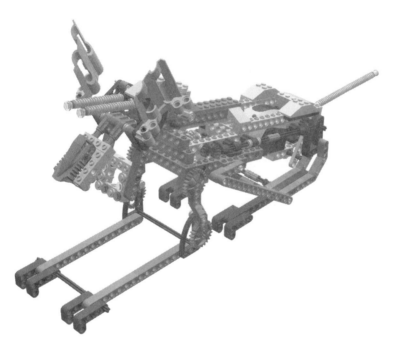

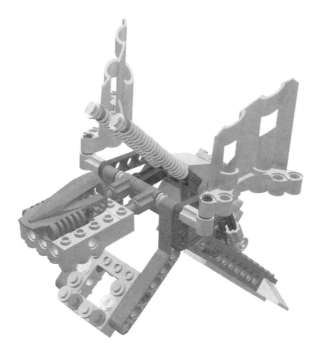

As Imperial troops spread across the galaxy, they often acquired native animals as pets. Over time they replaced these live prizes with their mechanical equivalents, and thus the Imperial Hounds were born. These Droids serve as both companions for the troops and as powerful tools of war on the battlefields of the Empire. They are often used to guard Imperial prisons and compounds, and also serve as scouts. Darth Vader has a pack of six Imperial Hounds that he can control with his thoughts, and the Emperor sometimes presents Imperial Hounds as gifts to officers who have distinguished themselves in the service of the Empire.

The Imperial Hound presented here will raise and lower itself on its back legs and wag its tail. Real Imperial Hounds are seldom in such a good mood!

Bricks & Chips...

Micro Scout Programs

The Imperial Hound will work with any of the Micro Scout's programs.

The Body

Begin by assembling the body sub-assembly of the Imperial Hound.

Body Step: Complete

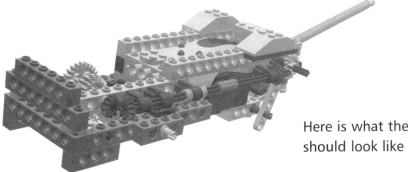

Here is what the body sub-assembly should look like when it is complete.

Body Step: 1

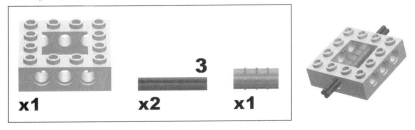

Body Step: 2

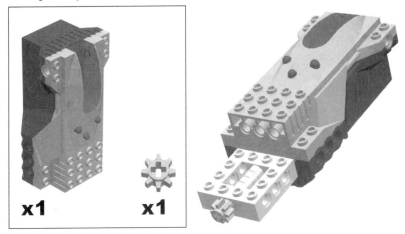

Body Step: 3

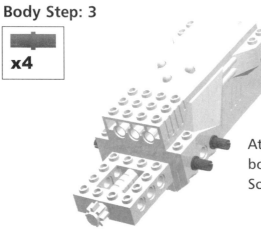

x4

Attach these four pegs to both sides of the Micro Scout as shown.

Body Step: 4

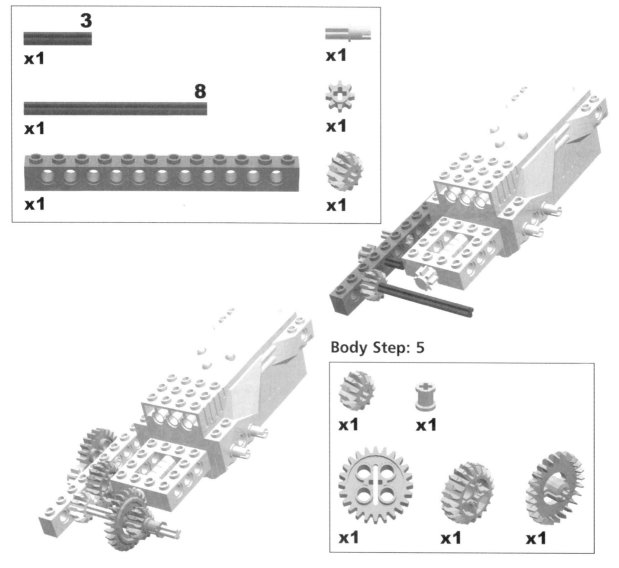

3

x1

8

x1

x1

x1

x1

x1

Body Step: 5

x1

x1

x1

x1

x1

Body Step: 6

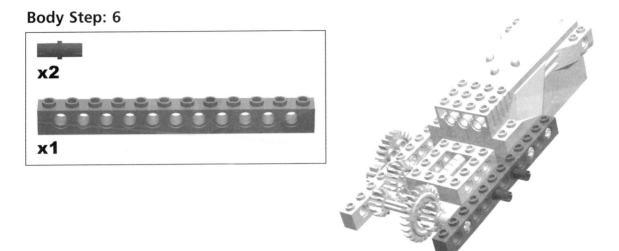

x2

x1

Body Step: 7

x1

x1

x1

x1

x1

x2

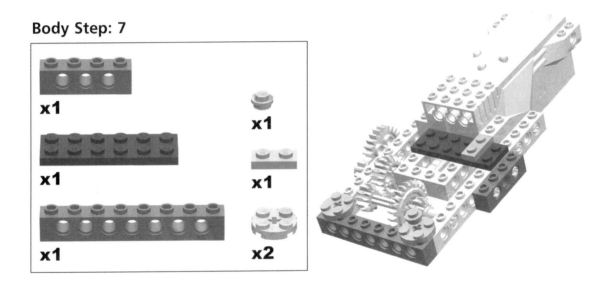

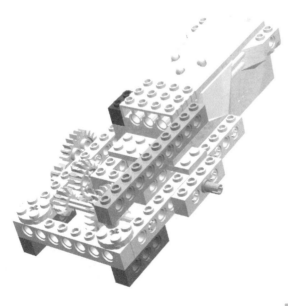

Body Step: 8

x2

x1

x1

x1

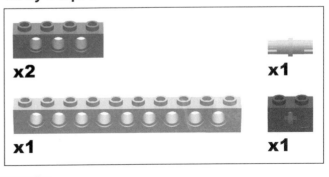

Body Step: 9

x1

x4

x1

3

x1

Body Step: 10

x1

x1

x1

x2

x4

x1

Body Step: 11

x2

10

x1

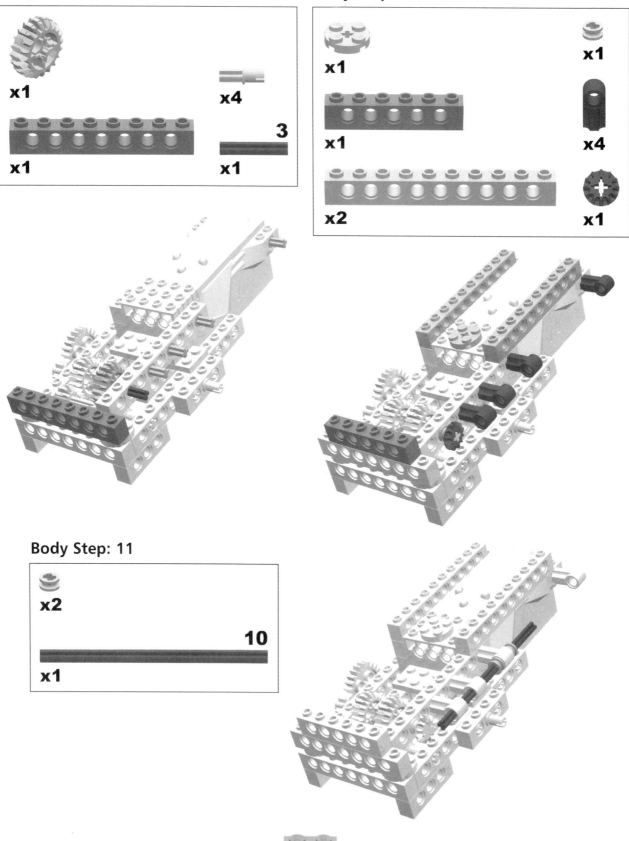

Body Step: 12

x1

x2

8

x1

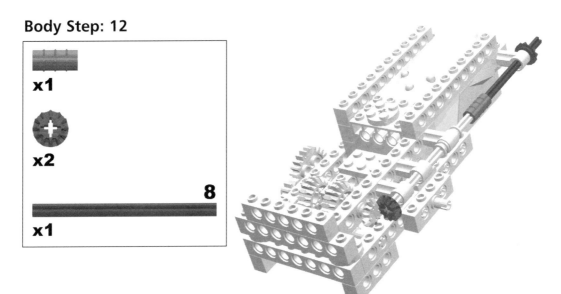

Body Step: 13

x1

x2

x1

x2

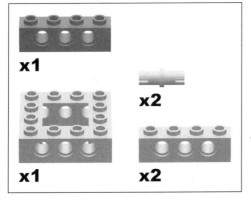

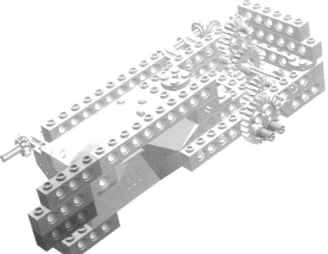

Body Step: 14

x2

x1

x1

x2

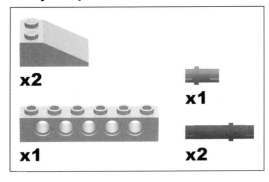

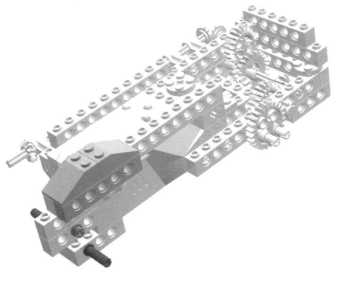

Body Step: 15

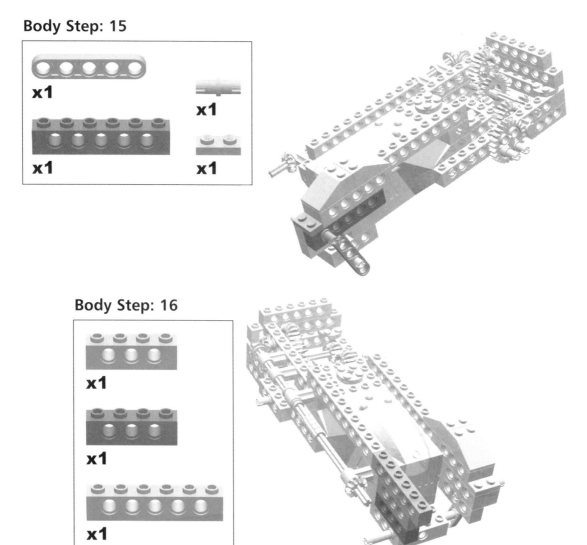

Body Step: 16

Body Step: 17

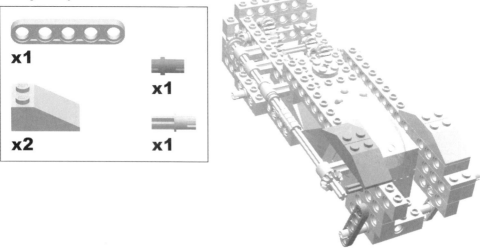

Body Step: 18

x1

x1

Body Step: 19

x1

5

x1

x1

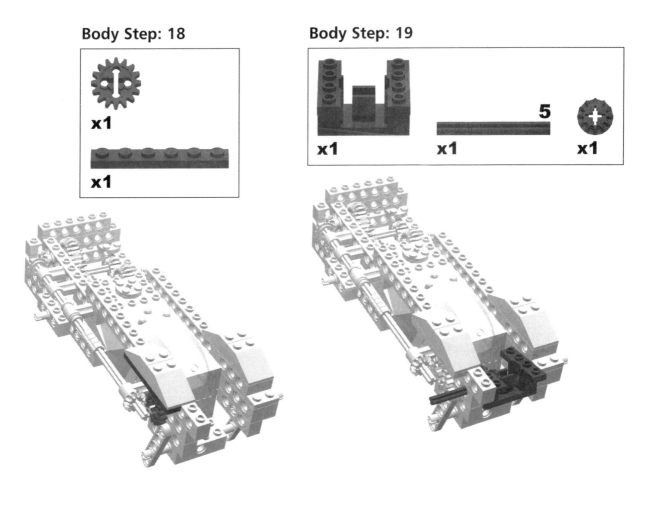

Body Step: 20

x1

x1

5

x1

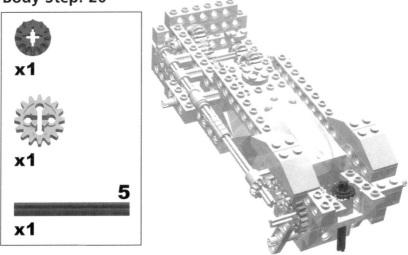

Body Step: 21

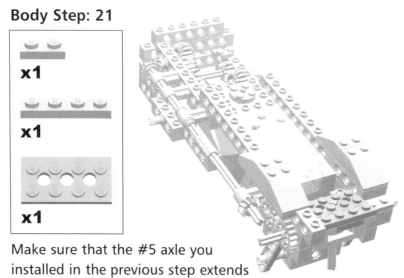

x1

x1

x1

Make sure that the #5 axle you installed in the previous step extends from the middle hole of the 2x4 plate as shown.

Body Step: 22

x1

x1

x2

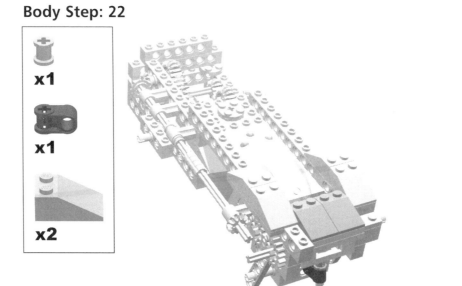

Body Step: 23

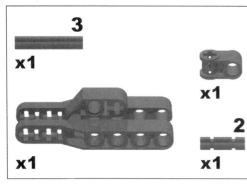

3

x1

x1

x1

2

x1

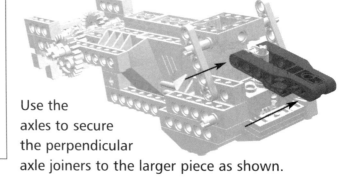

Use the axles to secure the perpendicular axle joiners to the larger piece as shown.

Body Step: 24

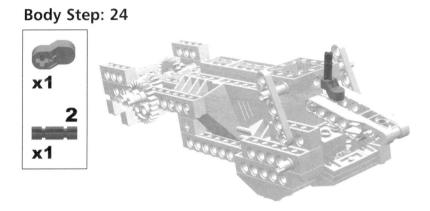

Body Step: 25

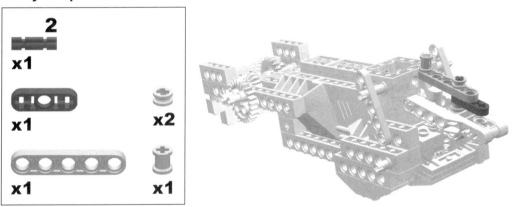

Body Step: 26

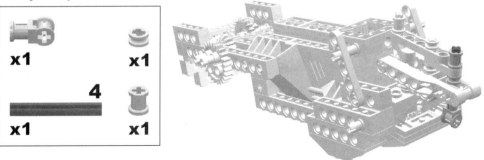

Body Step: 27

x1

17

x1

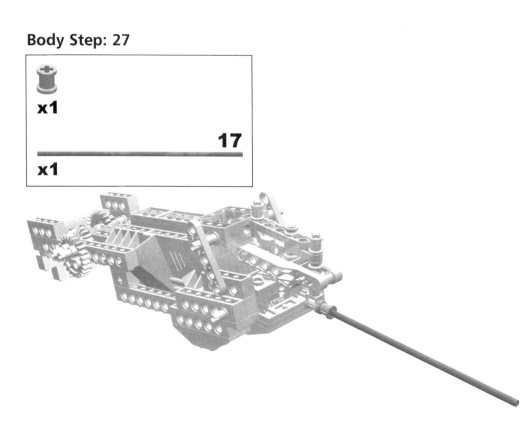

Body Step: 28

x1

x2

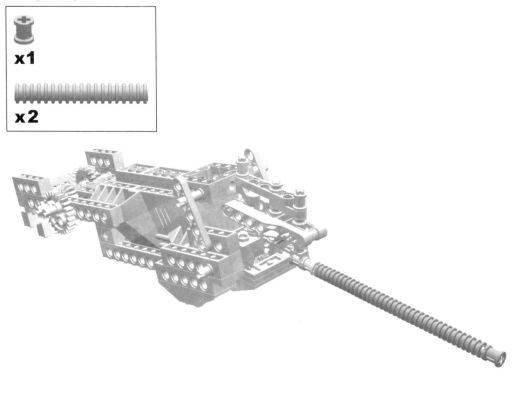

The Head

Head Step: Complete

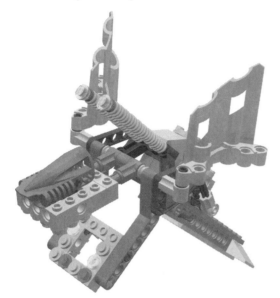

Here is what the head sub-assembly should look like when it is complete.

Customizing the Imperial Hound

Because the head and the front legs of the Imperial Hound are not attached to the Micro Scout, these are the two sub-assemblies that are the easiest to customize. You can easily turn the Imperial Hound into a bird, a dragon, or some other animal by building a new head for your robot.

Head Step: 1

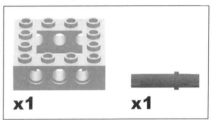

x1   **x1**

Head Step: 2

x1

x1

Head Step: 3

x2

2

x1

x1

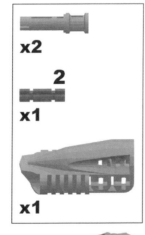

Place the nose-like piece as shown. To secure it in place, slide the #2 axle through the hole in the piece's side into the perpendicular axle joiner you added in the last step, as shown.

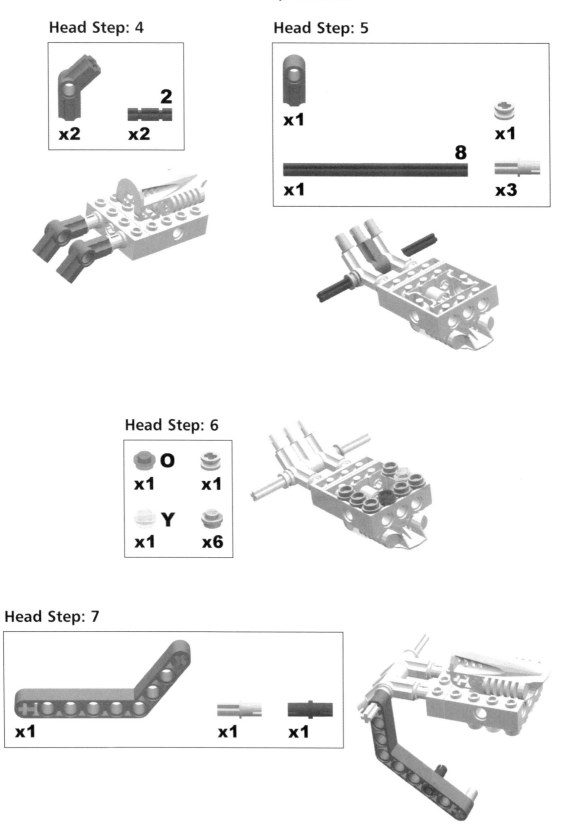

Head Step: 4

x2   2 x2

Head Step: 5

x1   x1

8 x1   x3

Head Step: 6

O x1   x1

Y x1   x6

Head Step: 7

x1   x1   x1

Head Step: 8

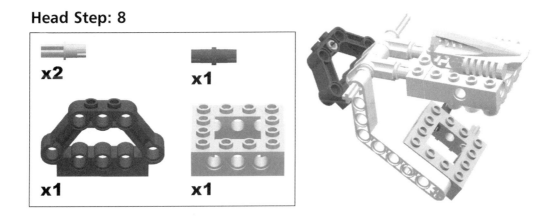

x2

x1

x1

x1

Head Step: 9

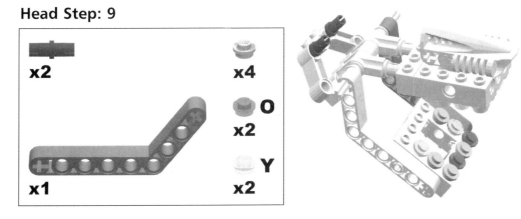

x2

x4

O
x2

Y
x2

x1

x2

Head Step: 10

8

x2

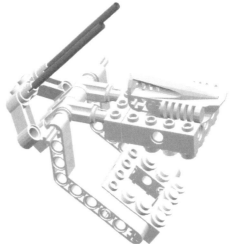

Head Step: 11

x2

Head Step: 12

x2

Head Step: 13

B
x2

Attach the eyes as shown.

Head Step: 14

x2   x2

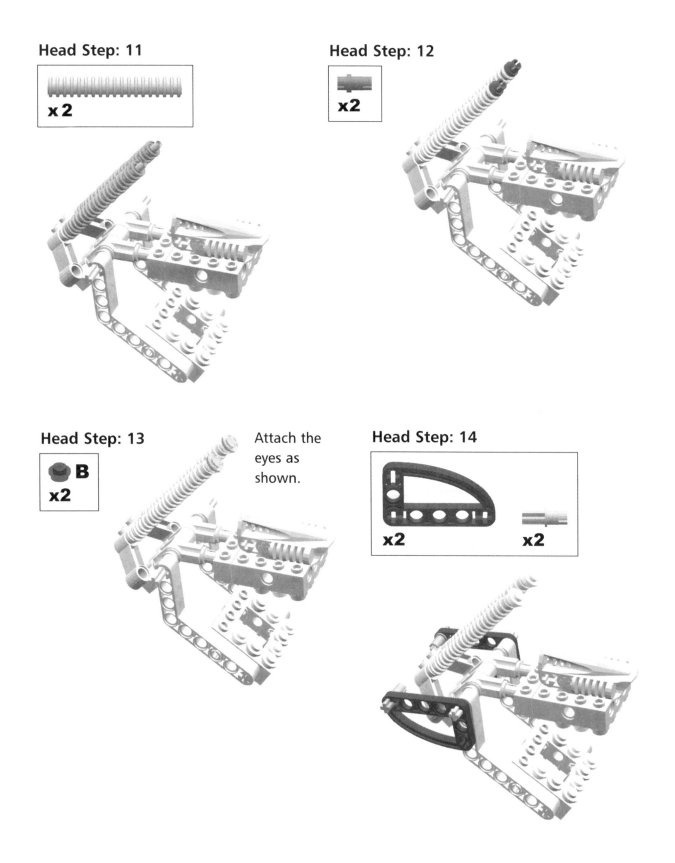

Head Step: 15

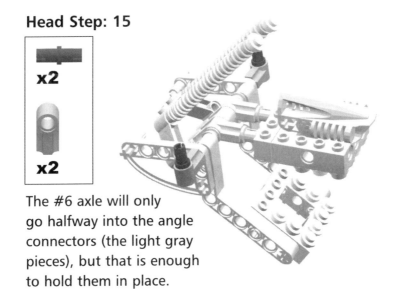

x2

x2

The #6 axle will only go halfway into the angle connectors (the light gray pieces), but that is enough to hold them in place.

Head Step: 16

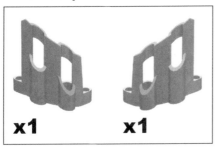

x1   x1

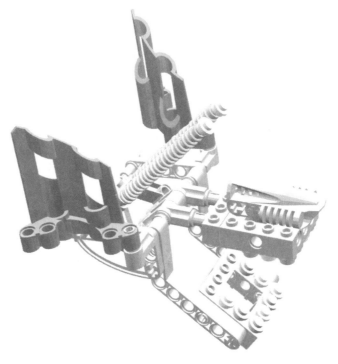

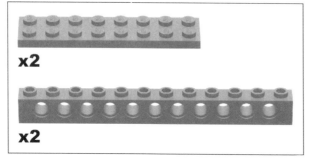

Head Step: 17

x1

x2

x2

Turn the head around and add these pieces as shown.

Head Step: 18

x2

x3

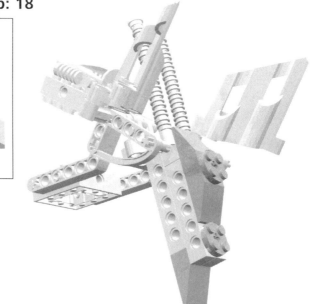

Head Step: 19

x2

x2

Head Step: 20

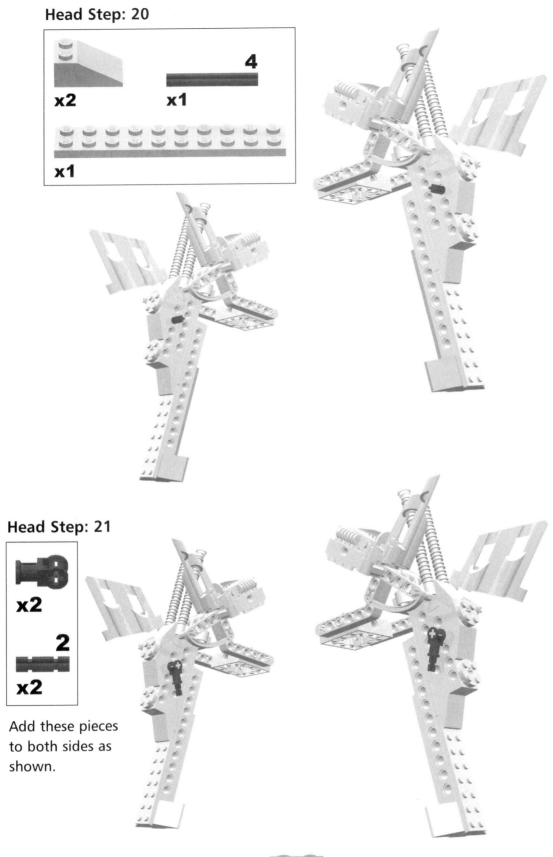

Head Step: 21

Add these pieces
to both sides as
shown.

Head Step: 22

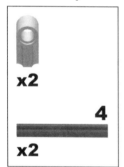

Add these pieces to both sides as shown.

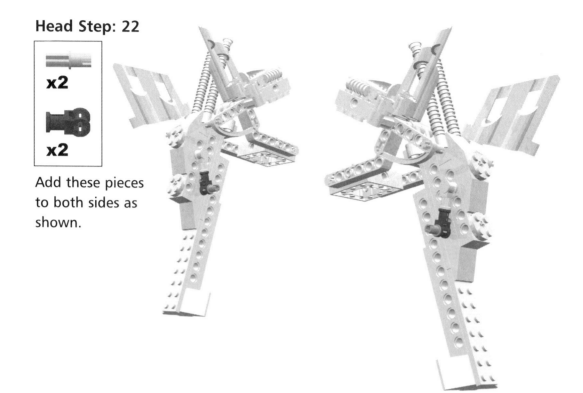

Head Step: 23

Add these pieces to both sides as shown.

Head Step: 24

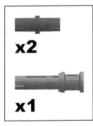

x2

Add these pieces to both sides as shown.

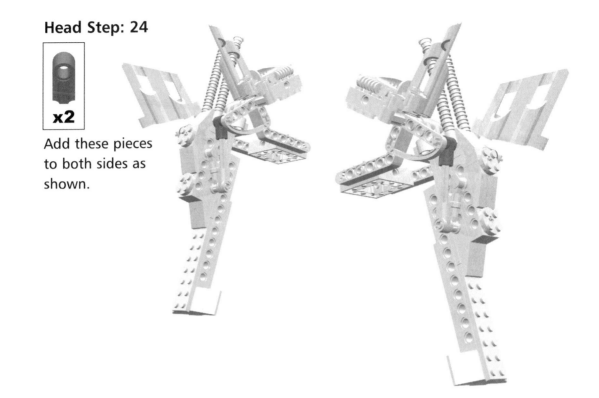

Head Step: 25

x2

x1

Add these pieces to both sides as shown.

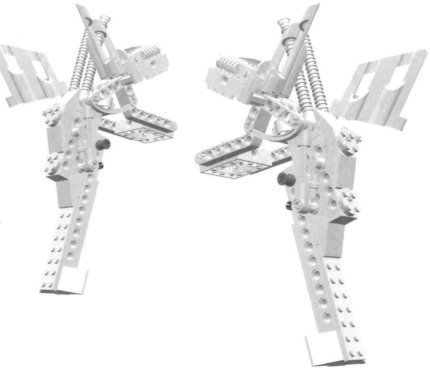

The Back Legs

The Imperial Hound raises and lowers itself on the back legs sub-assembly.

Back Legs Step: Complete

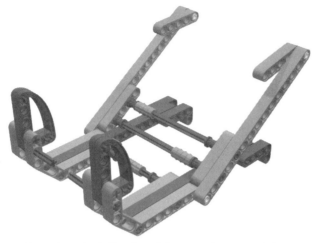

Here is what the back legs sub-assembly
should look like when it is complete.

Back Legs Step: 1

x4

x2

Back Legs Step: 2

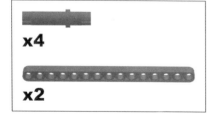

x4

x2

Back Legs Step: 3

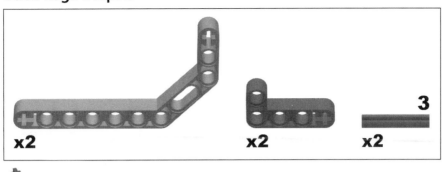

x2   x2   **3** x2

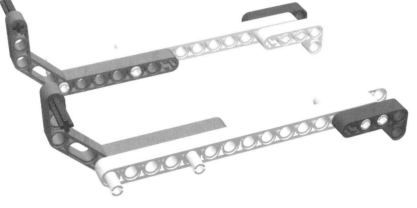

Back Legs Step: 4

x4   **5** x2   x2

Back Legs Step: 5

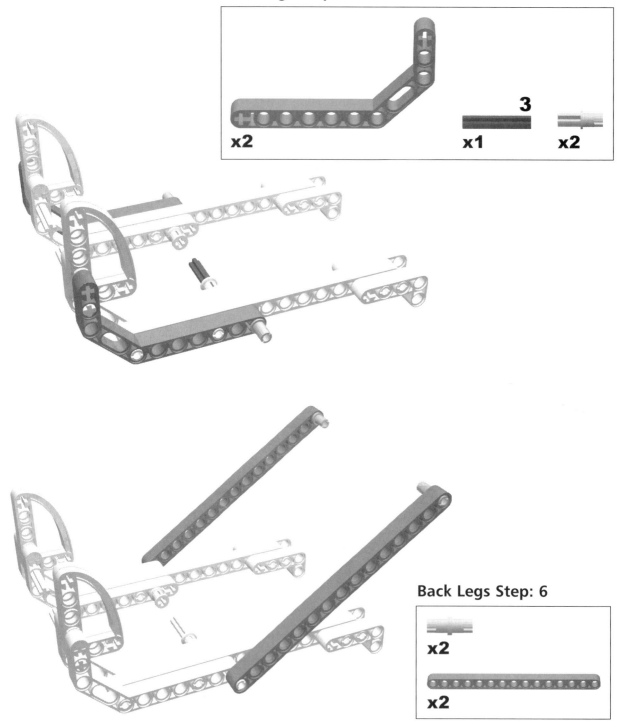

Back Legs Step: 6

Back Legs Step: 7

x4

2

x2

x4

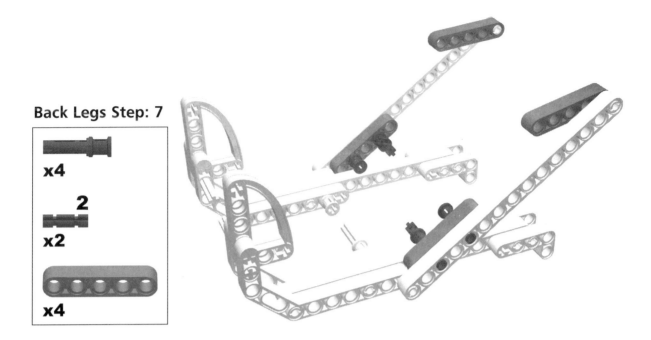

Back Legs Step: 8

x3

3

x1

8

x2

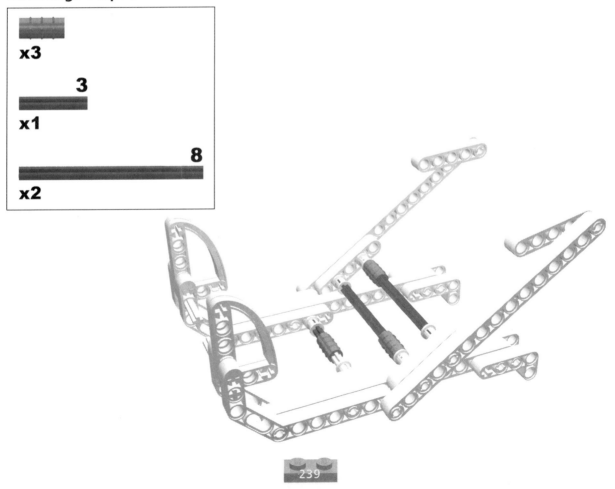

The Front Legs

The front legs of the Imperial Hound are purely decorative.

Front Legs Step: Complete

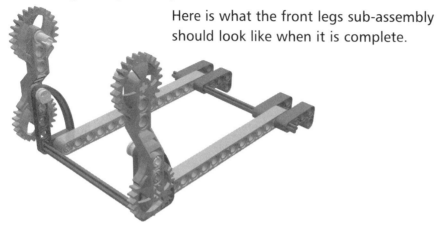

Here is what the front legs sub-assembly should look like when it is complete.

Front Legs Step: 1

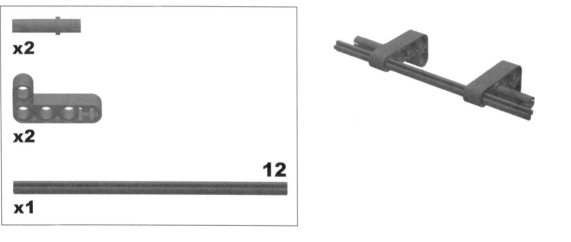

x2

x2

12

x1

Front Legs Step: 2

x2

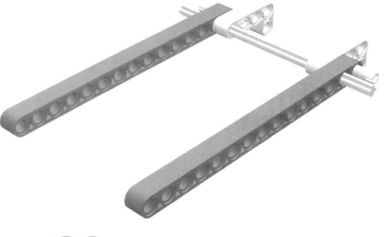

Front Legs Step: 3

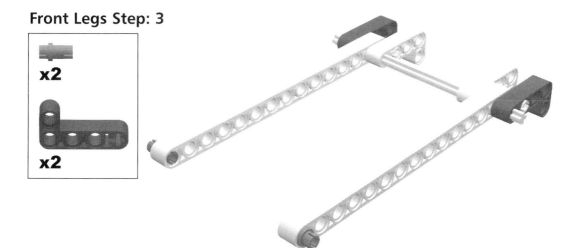

Front Legs Step: 4

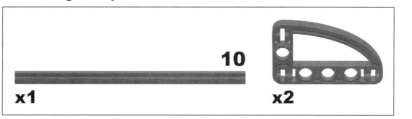

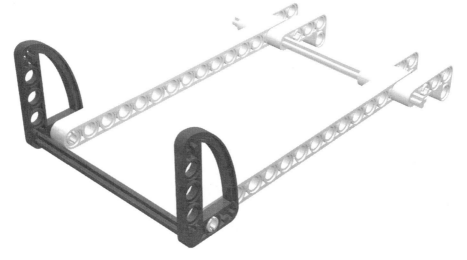

Front Legs Step: 5

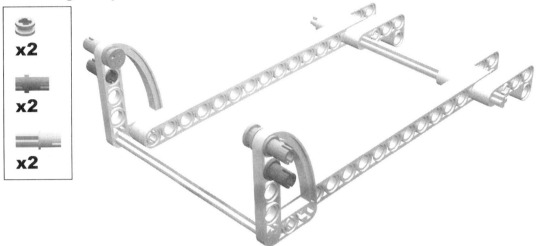

Front Legs Step: 6

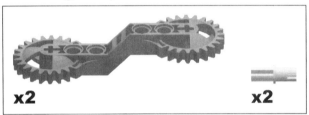

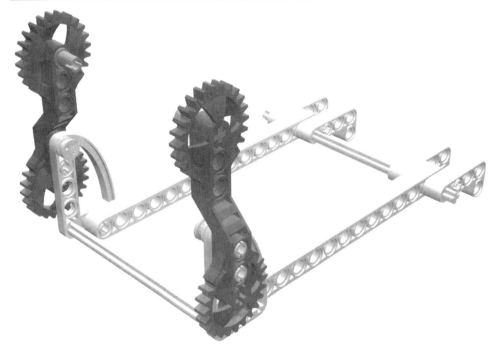

Final Assembly

You are now ready to take all of the sub-assemblies you have built (the front and back legs and the head) and attach them to the body to create the finished Imperial Hound.

Final Assembly Step: Complete

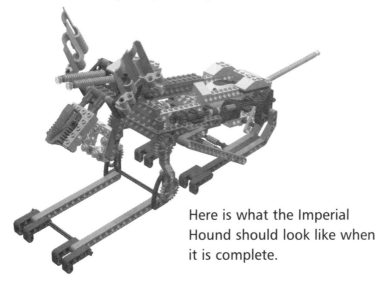

Here is what the Imperial Hound should look like when it is complete.

Final Assembly Step: 1

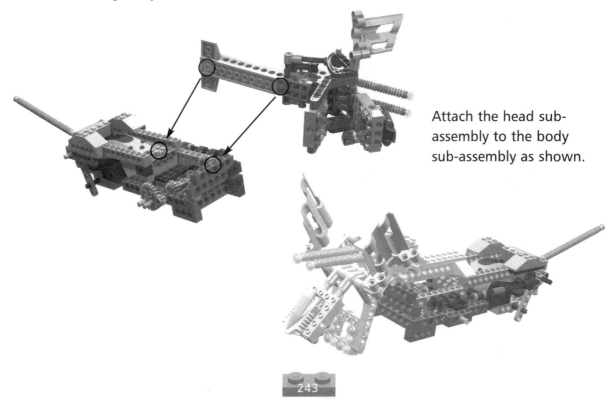

Attach the head sub-assembly to the body sub-assembly as shown.

Final Assembly Step: 2

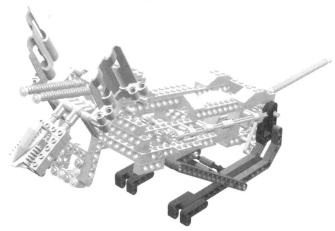

Attach the back legs sub-assembly to the body sub-assembly as shown. You will need to pull the back legs apart to attach them.

Final Assembly Step: 3

Attach the front legs sub-assembly to the body sub-assembly as shown and your Imperial Hound is complete!